Everywhere and Everywhen

Everywhere and Everywhen
Adventures in Physics and Philosophy

Nick Huggett

OXFORD
UNIVERSITY PRESS
2010

OXFORD
UNIVERSITY PRESS

Oxford University Press, Inc., publishes works that further
Oxford University's objective of excellence
in research, scholarship, and education.

Oxford   New York
Auckland   Cape Town   Dar es Salaam   Hong Kong   Karachi
Kuala Lumpur   Madrid   Melbourne   Mexico City   Nairobi
New Delhi   Shanghai   Taipei   Toronto

With offices in
Argentina   Austria   Brazil   Chile   Czech Republic   France   Greece
Guatemala   Hungary   Italy   Japan   Poland   Portugal   Singapore
South Korea   Switzerland   Thailand   Turkey   Ukraine   Vietnam

Copyright © 2010 by Oxford University Press, Inc.

Published by Oxford University Press, Inc.
198 Madison Avenue, New York, NY 10016

www.oup.com

Oxford is a registered trademark of Oxford University Press.

Library of Congress Cataloging-in-Publication Data
Huggett, Nick.
Everywhere and everywhen : adventures in physics and philosophy / Nick Huggett.
   p. cm.
ISBN 978-0-19-537951-8; 978-0-19-537950-1 (pbk.)
1. Physics—Philosophy.   I. Title.
QC6.H824   2009
530.01—dc22   2009016318

9 8 7 6 5 4 3 2

Printed in the United States of America
on acid-free paper

For my paradoxical twins, Kai and Ivor

Preface

I remember when I discovered that you could be a philosopher of physics. I was in the library of my school studying brochures for university admissions when I came across the Physics and Philosophy program at Oxford. It made perfect sense to me at the time; physics was my best subject and I'd developed an interest in philosophy in a fairly typical teenage intellectual way. So I applied and got in.

In hindsight, though, I'm not quite sure what I thought the philosophical study of physics was. I certainly didn't give a very good answer to that question in my admission interview! That was something I learned later, during my studies. (In fact, according to Rom Harré, a founder of the program at Oxford, the intention was not to produce philosophers of physics, but future leaders, well grounded in reasoning, ethics, and the sciences.) Still, I found that my youthful intuition was reliable, and after I completed my undergraduate studies I went on to Rutgers University in New Jersey, where I was lucky enough to work with some of the best and most generous philosophers of physics I have ever met. Best of all, afterward I found a job where I could research and teach my subject, at the University of Illinois in Chicago.

So now my only problem is explaining to people—parents, friends, neighbors on planes and at dinner parties, and especially physicists and philosophers—what it is that I do. Something to do with the ethics of science? That's an important topic, but generally not part of philosophy of physics. Or perhaps the connections between Buddhism and quantum physics? That idea was popularized in the 1970s by Fritjof Capra and Gary Zukav, but it's not what most philosophers of physics are interested in. Or is it an attempt to tell physicists what must be the case by pure speculation, regardless of the facts of experiment? Or perhaps to show that physics is nothing but a social fabrication, that truth about the physical world is not objective but whatever physicists decide. Well, the sociological dimensions of science are interesting, and some people do take a very hard line, but most philosophers of physics take very seriously physics' ability to get at objective truth—and they think that it is the ultimate standard, not pure speculation.

So this book is to explain to all those people some of the ways in which physics and philosophy can be in fruitful dialogue—it is that dialogue that engages philosophers of physics (and, as we'll see, many physicists). We shall see that indeed the traffic is two-way: while physics has important lessons for philosophy, the kind of investigations at which philosophers excel are necessary in science, and some of the most important advances in physics have required philosophical contemplation. To be more specific, we will see three main kinds of interaction.

First, there are cases in which philosophical questions can be formulated in a precise way in physics and then addressed with resources of physics. For instance, could space have an edge? Second, there are cases in which ideas used in physics turn out to be conceptually unclear or incompatible with new knowledge in physics. What physics requires here is a careful analysis of the concepts and an understanding of how they are used. That kind of work is philosophical, though it is often done by physicists—'philosopher-scientists', as Einstein was described. For instance, what is it for events to be "simultaneous"? And third and finally, the fact that we are physical beings living in a physical world of a specific type has profound consequences for the way we experience the world. Having an understanding of these consequences is crucial for a clear philosophical view of a variety of problems. How, for instance, do we perceive left *versus* right handedness?

To see these things in more detail, naturally we'll need to introduce some physics and philosophy. You'll notice some difference between the two here. The physics will be presented largely as the materials for our discussion, while I will be showing you how to think about the physics like a philosopher. When you read a popular book on physics, the goal usually is to explain recent developments in terms that are accessible to the layperson; the details themselves take years of study even for very smart people. The best books of this kind do a great job of explicating the fundamental ideas and implications, but of course they don't make you a physicist. I have a rather different ambition for this book. This is not a book that just seeks to explain recent developments in philosophy of physics—though we will talk about some of them—but one that aims to help the reader really think philosophically about physics and the physical world. Having taken ten years in higher education to become a philosopher, I hesitate to say that this book will make the reader a philosopher of physics, but I do hope it will show the way and allow a first step in that direction. To put it another way, the book doesn't just report on philosophy, it does it too, and I hope that example will be useful.

As a result this book will be demanding in places. Philosophy involves patient reasoning, canvassing of different possible positions, step-by-step argument, and to-and-fro. Sometimes it takes effort to keep the logic of the topic clear. However, I've picked pieces of physics and of philosophy that are suitable for a general audience (I've taught these topics to a lot of undergraduates of very different abilities, so I have a pretty good sense of what is digestible). The bottom line is that I've picked topics that should be fully comprehensible to anyone who is prepared to apply some careful thought.

I offer the following deal: in return for carefully thinking through some sometimes challenging (but always interesting, I hope!) arguments, the reader will start to learn to be a philosopher of physics.

Here's the plan of action. In the first chapter I will give an example of a philosophical problem so that the reader can see right away what kinds of concerns and what kinds of approaches drive the book. We'll also fill in some of the physics background that we need, and some important philosophical concepts (just what is a 'law' of physics?). Chapters 2–3 discuss Zeno's paradoxes, which challenge the very idea of motion. For instance, an arrow cannot move during any instant because an instant has no duration. But if it doesn't move at any time then how does it move at all?

Chapters 4–6 concern the overall 'shape' of space, for instance whether it has an edge, whether it is 'closed up' on itself, and whether it has more than three dimensions, and if not why? Chapters 7–8 continue the discussion of the shape of space in a different direction, investigating its geometry. Is it flat? What would it mean if it weren't? How could we tell? And generally, what does it mean to say space is curved? Chapter 9 completes the discussion of space by asking the obvious remaining question–what is it?

Chapters 10–11 take up the issue of time. Time seems so different from space: for instance, we certainly experience time differently than we do space, as something 'moving'. How could physics account for this fact? Chapters 12–13 are devoted to another puzzling aspect of time: is time travel possible? We'll see possibilities and restrictions, and see ultimately how it is a coherent possibility.

Chapters 14–15 explain and investigate Einstein's relativity, which crucially changes our understanding of space and time. The presentation is a little more rigorous than most popular presentations, but all that is involved is a simple, if unfamiliar, geometry. We will be able to show why moving bodies contract and why moving clocks slow down, and understand what this means.

Chapters 16–18 are devoted to some issues that have been of particular interest to me. First is the question of what it is for an object to be left rather than right; what is it about a left hand, or left-handed glove, or left-handed screw that makes it left rather than right? After all, all these things are very, very similar to their mirror images. And then there is the question of identity and indistinguishability in physics. The particles of physics are exactly alike, so does it make any difference if they swap their locations, say? Are they like money in the bank or are they like people?

A note on citations. To maintain an informal style I have gathered annotated references at the end of each chapter instead of inserting citations in the main text. I have also omitted certain more technical works, while crediting authors. I hope they will understand my aims and forgive me.

I've had a lot of help with this book, first and foremost from all my students who have read it or discussed the issues. Among them I should especially thank my research assistants David Lee, Poonam Merai, and Darrell Wu, and my TAs Rashi Agarwal (especially for 'Mr. Toody'), Maria Balcells (especially for 'spatter'), Isaac Thotz, and Aleks Zarnitsyn. I also had feedback and assistance from a number of friends and colleagues, particularly Craig Callender, Stanley Fish, David Hilbert, Rachel Hilbert, Tom Imbo, Jon Jarrett, David Malament, Tim Maudlin, Chris Wüthrich, and Eric Zaslow. Because of the informal style, I have not credited anyone with specific contributions; they know how they helped, and I hope that suffices.

Of these I wish to thank Tim Maudlin especially, for teaching me much of what I know of these topics in the first place, and for extremely generous and helpful comments on the manuscript. His contributions are too numerous to credit individually, but they certainly made this a much better book.

With such assistance, I can sincerely say that any remaining faults in this book are mine alone. (And of course, none of the above necessarily agree with my arguments, or their conclusions.)

Chapters 17–18 are based closely on an unpublished essay that I wrote with Tom Imbo, itself based on research carried out by him and his graduate students Randall Espinoza and Mirza Satriawan. I am very grateful for all I have learned from Tom during the time we have worked together, and for his permission to use this work here.

I am also grateful for the support I have received from Oxford University Press. Peter Ohlin has been very encouraging, and Stephanie Attia did a great job polishing the text.

I started writing this book in the hospital when my children were born, as a way of keeping my hand in while living with infant twins, so thanks to them for concentrating my mind. And of course thanks to my wife, Joanna, for her enormous patience and selfless support of all I do. Last of all I want to thank my father, Cliff; it was talking to him about my work that got me started on this book, and in addition to discussing the book with me he has served as the model of my intended audience.

Contents

Everywhere and Everywhen

1

A Longish Introduction

The Problem of Change

In this chapter I'll lay out a few crucial ideas from physics and philosophy that we will use in later chapters. We need to know some key points of Newton's physics, for instance, and we need to understand what it means for something to be a 'law'. Rather than discuss these things laundry-list style, I'll introduce them in the context of a brief history of an important philosophical question—perhaps the original philosophical question—'what is change?' In this way we can see right away how a philosophical inquiry into physics works.

A number of the philosophical issues that we will discuss concern the nature of change and difference—how is change mathematically possible? Would anything be changed if the world had a different shape? If the world were relocated in space? Why do we experience change over time so differently from variation in space? We will see how some of these questions lead to advances in physics and mathematics and how some require revisions in our assumptions about the world at the most fundamental level. The first argument we will discuss is supposed to show that change is not possible at all.

This rather strange sounding conclusion was first found in the writings of Parmenides, a Greek philosopher from the early fifth century B.C. Such arguments are obviously not from experience, since things do appear to change; the argument is one from logic, and aims to show that experience is misleading. (Of course it sometimes is: you think you recognize someone across the street, but when you get closer it isn't them after all.) In chapter 2 we will discuss at length the most powerful and important arguments against change, those of Parmenides' pupil Zeno. First we will consider the argument offered in support of Parmenides by another Greek philosopher, Melissus, who worked in the mid-fifth century B.C.

1.1 MELISSUS'S PARADOX

Let's start with Melissus's words:

> And it cannot perish, or become greater, or be rearranged, or feel pain or
> distress. For if it experienced any of these, it would no longer be one. For
> if it became different, it is necessary that what is is not alike, but what
> previously was perishes, and what is not comes to be.

It's hard to be sure that one is getting 2,500-year-old scraps of writing right, but here is a reasonable stab at Melissus's reasoning. At the
beginning and end of a change, the thing that changes must be both
different—so that change has occurred—and the same—so that something has changed.

For instance, a tree turning from green to brown in the fall differs at
the beginning and the end of the change, but it must be one and the
same tree; otherwise we don't have something changing, but rather two
different trees. More abstractly, if X is changing and t_1 and t_2 are the start
and end times of the change, X at t_1 both is (since it is the same as) and is
not (since it is different from) X at t_2. But that is to say that any change
requires both difference and sameness, which is a contradiction and hence
is impossible.

Well, surely this argument is based on some kind of confusion or trick,
but what exactly? In his *Physics*, Aristotle (382–322 B.C.) showed how
the reasoning was based on a conflation of two senses of the word 'is'.
(President Clinton infamously made a similar point about 'is' to a Grand
Jury during the Lewinsky scandal, though he was making a distinction to
do with tense.) If we spell out Melissus's argument a little more carefully
we can see how a conflation is involved.

Suppose it can be truly said that one thing is another, 'W is X': for
instance, that Peter Parker is Spiderman. And further suppose that Y is
Z: for instance, Bruce Wayne is Batman. And suppose finally that that X
is *not* Z. Then we can conclude that W is not Y: since Spiderman is not
Batman, we can conclude that Peter Parker is not Bruce Wayne.

The same line of thought lies behind Melissus's argument. The tree *is*
green at the start but *is* brown at the end, but green *is not* brown. So,
it seems, just as Peter Parker is not Bruce Wayne, the tree at the start *is*
not the tree at the end. But the tree at the start *is* the tree at the end,
because the change is to a particular tree. Therefore the change involves
a contradiction and is thus impossible.

But putting it this way, we can see the fallacy. The argument about
Spiderman and Batman works because the sense of 'is' all the way through
is that of 'is the very same one as': the so-called is of identity. (We denote
this '=' in math: $2 + 2$ 'is the same number as' 4.) But in Melissus's
argument when we say the tree *is* green, we are not saying that it is the
very same thing as the color green (whatever that would mean). Instead
we are using 'is' to ascribe a property; this sense is the 'is of predication',
and here it predicates greenness. Thus we cannot reason as we did for the

superheroes and thereby conclude that there are different trees; hence Melissus's contradiction does not arise.

That's not to say that there are no remaining problems. Aristotle's picture of change is of an underlying thing that takes on different properties at different times. But how does it make sense for a tree to be green and brown, even at different times? Surely nothing can be green (all over) and brown (all over). So do we mean that there are different properties for each time: the tree is green-in-September but not green-in-October. But isn't green just green? So maybe, just as the tree is made of leaves, roots, trunk, and branches, it is made of different temporal parts: the part in September is green and the part in October is brown. But then we don't have change in Melissus's sense after all: the 'change' merely involves two distinct parts of different colors.

Well, we won't pursue these issues further here (they will be the background of our discussion of time). For now, we have seen how Melissus issued a philosophical challenge to the most basic concept of physics, that of change; and we have seen how the challenge was met, not with a modification of physics, but with an advance in the understanding of language. This lesson is one that philosophy has learned many times; sometimes problems arise just from a confusion about the meanings of words.

1.2 WHAT IS CHANGE?

At the most general level, something changes when one of its properties (green, say) is replaced with another (brown, say). But is there more to be said about what happens in any change? As we'll see, there is, and it makes a big difference to the way we understand the world—to the kind of scientific theories we accept. Indeed, we will see that the conception of change has itself evolved, with dramatic implications for scientific progress.

Aristotle

After defending change, Aristotle went on to explain what it was in more detail. The key point is his belief in *formal* or 'natural' explanations. Aristotle's notion of nature is broader than ours. For instance, he offers here the explanation of musical octaves as an example: a note in one octave corresponds to the one of double the frequency in the octave above, to that of half the frequency in the one below. In this case it is the existence of the sequence of numbers (thus frequencies) obtained by repeated doubling of a number that formally explains octaves: it is in the nature of things that there are such numbers.

In fact Aristotle is appealing to Plato's theory of forms: that everything we perceive is merely a flawed copy of some perfect archetype or 'form'— the sound of the flute as a replica of the actual sequence of doubling numbers. (Plato famously allegorizes us to people chained in a cave and seeing only the shadows of things outside cast on the wall of the cave. The things are the forms, and the shadows the copies.) For Aristotle, however, the forms are not otherworldly ideals, but express how things *should* be.

Thus his forms enter into teleological explanations, those in terms of '*ends*'. To use his illustrations, why do spiders spin webs and plants grow roots in the soil? In order to get food, of course, and thereby thrive as the kind of organisms they are—to achieve their ideal state or 'form'. Or to give another example, which we shall draw on repeatedly to develop some important ideas, why do some things appear to orbit the Earth while others fall when dropped?

According to Aristotle, the universe is a sphere, with the Earth (itself spherical) at the center. Just inside the surface of the universe are the stars, in daily rotation about the center—and hence the Earth. Then moon, sun, and visible planets move daily with the stars but also orbit the center slowly, and hence appear to move slowly through the stars from day to day. (See figure 1.1.)

Why do the stars and planets (Aristotle included the sun and moon among the planets, since they move relative to the 'fixed' stars) have such motions? Because, according to Aristotle, they are made of the 'fifth element', '*aether*', whose nature and form are circular motion about the center. That is the ideal state of the aether, and so its nature causes it to move circularly. (Actually Aristotle seems uncharacteristically confused here, because the stars don't all move around the center, but around the Earth's axis.)

Moreover, the forms of the other four elements—earth, water, air, and fire—are a specific location in the universe. For instance, it is in the nature of earth to be at the center of the universe. Thus any earth lifted and released will naturally move in order to realize its form, causing it to fall, explaining the heaviness or 'gravity' of earth (and why the planet Earth is at the center). More generally, anything that contains a considerable amount of the element earth will also fall. So once again, things are caused to change in a certain way in order to attain the goal of a natural end. (The natural places of water, air, and fire are concentric shells around the center in that order, and so they will move naturally to those places.)

Descartes

In the seventeenth century the 'scientific revolution' not only developed new scientific knowledge, it introduced a whole new view of science itself, including conceptions of the nature and causes of change. Those involved were very critical of Aristotle's understanding of change, particularly criticizing it for offering explanations too cheaply—*anything* can be explained by attributing suitable natures. Isn't saying that rocks fall because it is in their nature to be at the center just to say that they fall because they have the power of downward motion? As Molière joked, isn't that explanation

Figure 1.1 In Aristotle's spherical universe the Earth is at the center, surrounded by the planets and stars. (Image courtesy History of Science Collections, University of Oklahoma Libraries; copyright the board of Regents of the University of Oklahoma).

as informative as saying that opiates cause sleepiness because inducing sleep is their nature. In other words, not informative at all.

Likely these complaints are not entirely fair to Aristotle—for instance, he did not believe that everything that animals habitually do could be explained by their natures—and they are better targeted at later followers. But the founders of modern science were convinced that a better conception was needed: one such was the 'mechanical view', which was most fully articulated by René Descartes (1596–1650).

We will discuss Descartes's views on the nature of matter further in chapter 9, but for now the important features are that he believed: (1) the universe is completely *full* of matter, and (2) all matter is essentially the *same*, (3) with no fundamental properties except *size, shape*, and *relative position*. These properties are the basic geometrical ones, so the universe that Descartes envisions is one in which there is geometrical body at every place. Add to that the dimension of time, and we have a

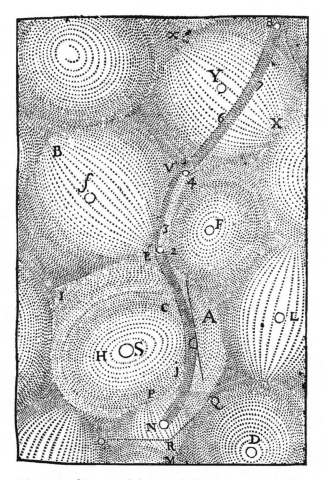

Figure 1.2 A part of Descartes's 'geometrical' universe. The circles represent celestial bodies: for instance, S is the sun, and the 'cell' around it our solar system. The universe is completely full of matter, and the dotted lines show how bodies are moving at various locations. For instance, there is a huge 'vortex' of matter swirling around the sun, carrying the planets with it, which represents a comet moving from star to star: Halley did not discover that comets orbit the sun until 1705. (Courtesy of Special Collections and University Archives, University of Illinois at Chicago Library.)

world of geometrical bodies in motion and changing shape and size; see figure 1.2.

However, we certainly perceive more properties in things than geometric ones: not only are bodies also sometimes red or hot, they also can be capable of growth, or be heavy, or be able to put people to sleep, and so on. According to Descartes, *all* other properties arise solely from the fundamental geometrical properties of bodies:

> Therefore, all the matter in the whole universe is of one and the same kind; since all matter is identified solely by the fact that it is extended. Moreover, all the properties which we clearly perceive in it are reducible to the sole fact that it is divisible and its parts movable. . . . I know of no kind of material substance other than that which can be divided, shaped and moved in every possible way, and which Geometers call quantity and take as the object of their demonstrations. And that there is absolutely nothing to investigate about this substance except those divisions, shapes and movements.

That is, all other properties of a body arise from the particular shapes, arrangements, and motions of its parts. Further, the only thing that can change the motions of bodies are collisions with other bodies, and of course the results of a collision are again determined by the geometric properties of bodies, especially their sizes and motions. Therefore, all the properties and all the changes of physical things are, according to Descartes's mechanical philosophy, at root geometrical. In contrast to Aristotle, this mechanical account of change is 'reductive': all features of the physical world are to be reduced to the geometric arrangements of bodies, without any forms or natures.

For instance, in this scheme the motion of the planets is explained in a wholly different way. According to Descartes, the matter that fills the solar system is in collective rotation about the sun and pushes the planets around with it; see figure 1.2 again. The solar system is of course full of matter because the universe is. It is not opaque because it is the medium through which light travels, in the form of pressure waves—a mechanical account of light.

Finally, there's also a smaller vortex around the Earth. Like all spinning things, it tends to move away from the axis of rotation (we'll talk about this idea more in chapter 9, but think of how coffee spills over the sides of a cup if you stir it too fast). According to Descartes, the vortex is composed of very fine bodies and so has a greater tendency to recede than the ordinary-sized bodies on the Earth. Thus in the 'competition' to move away from the Earth, the vortex wins and terrestrial objects are pushed down, explaining their weight mechanically.

Again, the point is to see that Descartes offers a very different kind of explanation than does Aristotle, and in general understands change in a very different way: not as resulting from forms, but as the result of geometrical changes.

Newton

The problem with Cartesian mechanism is that, despite the application of some of the greatest minds in physics, no one ever discovered a successful

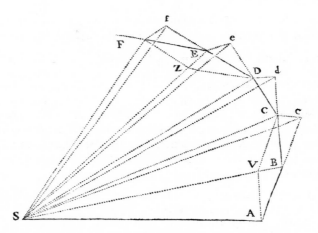

Figure 1.3 Newton used this diagram to calculate the motion of a body that was regularly pushed toward a center point S. The line ABCDEF represents the motion, and AS, BS, . . . , FS show the directions and locations of the forces (the other lines are part of Newton's calculation). We see—assuming that the forces continue—that the result is a path around S. (Courtesy of Special Collections and University Archives, University of Illinois at Chicago Library.)

mechanical account of the motions of the planets. The Cartesians were unable to turn Descartes's qualitative model into an empirically successful quantitative theory. Isaac Newton (1642–1727), however, succeeded by introducing the law of universal gravitation, according to which any two bodies exert a force on each other. (Quantitatively, the force is proportional to their masses, and inversely proportional to the square of the distance between them: double either mass and the force doubles; but double the distance between them and the force goes down by a factor of 1/4.)

The law gives an immediate explanation of why a body drops when lifted and released: there is a force between it and the Earth. But let's consider why the same law also explains the motions of the planets. First, as Descartes realized, if nothing happens to a body, then it will keep moving at a constant speed in a straight line. Now what would happen if a force was exerted on a body at regular intervals toward a center point? In between it would move in a straight line, but every so often its motion would be deflected toward the center: see figure 1.3.

As you see, if the forces are the right strength then the result is an orbit around the center (otherwise the body may just be deflected or ultimately reach the center). And of course, as the time between the collisions gets smaller, the orbit gets smoother; if the force acts all the time, the orbit is an ellipse. The result is general, a body orbits a center if there is a force toward that center; in other words, it constantly 'falls toward' it, instead of moving away on a straight line.

So both the weight of terrestrial bodies and the motions of the planets are explained by Newton in terms of the law of gravity. Unlike either Aristotle or Descartes, Newton identified them as examples of the very same phenomenon. How did he do this?

From the motion of the moon he could calculate how fast it was 'falling toward' the Earth away from a straight path. But the distance to the moon is about 15 times the radius of the Earth, so he knew from the law of gravity that the force on the moon would be 15^2 times stronger if it were at the Earth's surface. Therefore it would accelerate $15^2 = 225$ times faster at the Earth's surface—which, he calculated, was the very speed that all bodies actually do drop! That is, he compared the fall of the moon with the fall of objects on the Earth and found they were the same thing. This comparison is amazing even today, so Newton's insight in seeing as the same, two things that previously were considered completely different, makes this argument one of the most profound insights in the history of physics.

Now, according to Descartes's mechanical view, the force of gravity would have to be the result of collisions somehow, but Newton was unable to give a mechanical explanation of them and maintained a public stand of agnosticism about their causes. The Cartesians therefore argued that Newtonian gravity was a nonmechanical power; that the Newtonian picture of change is one in which bodies have Descartes's geometric properties *plus* nongeometric powers. (These now include not only gravity, but electromagnetic attraction and repulsion in virtue of electrical charges, and nuclear forces in virtue of 'nuclear charges'.) But Newtonian physics does not face the problem of cheap explanations, because the list of nongeometrical properties is not open-ended but strictly circumscribed. We cannot go on arbitrarily adding properties to explain how things are and change, but must make do with just a few.

This Newtonian view is the underlying picture in physics to this day, even in string theory (discussed later), which seeks a radically new understanding of the fundamental properties. (Well, quantum mechanics, discussed below, seems to require that something else be added to the Newtonian picture. However, it certainly gives a satisfactory image of physics for our purposes.)

1.3 LAWS

A sketch of Newton's physics is all we'll need for most of this book; presently we will introduce some more contemporary ideas that will come up at some points, but this section is primarily about the important concept of a *law* of physics. We already mentioned the law of gravity, but consider, for example, the other laws that comprise Newtonian mechanics (all of them are needed to really calculate the motions of the planets):

1. A body's acceleration is proportional to the force on it and inversely proportional to its mass: more force means more acceleration, but more mass means less.
2. As a special case, if there is no force on a body, then it does not accelerate at all and thus moves at a constant speed in a straight line.
3. Whatever force body *A* exerts on *B*, *B* exerts the same force on *A.* (So not only are you attracted to the Earth, the Earth is *equally* attracted to you; however, since it is much heavier, the effect on the Earth is much smaller. The same goes for the Earth and the sun.) For short, 'action = reaction'.

We ought to explain here the formal concepts of speed and acceleration, since they will be assumed in the book. So consider a moving object—a flying toaster, say. Suppose that during a minute it travels 15 miles (it's a fast toaster). Then the average 'speed'—the 'rate' at which distance is covered—during that minute is $15 \div 1$ miles/minute; the speed over any interval is the distance traveled divided by the length of the interval. (In chapter 3 we'll discuss speed at an instant–over zero seconds.) What if it changes its speed? Suppose that during the next minute the toaster flies 20 miles, for an average speed of 20 miles/minute? Then the rate at which it has increased its speed—'accelerated'—is $20 - 15 = 5$ miles/minute. (More accurately, speed *in a given direction*—'velocity'—is the relevant concept; in Newton's laws, acceleration is the measure of how fast velocity changes.)

What Is a Law?

What does it mean to call 1–3 'laws'? They're supposed to be true of course, but they are more than that. (In fact Newton's laws have been superseded, but they are 'true enough' for most of the solar system.) For instance, it's also true that there are eight planets (and some dwarf planets); is that a law? No, it's just something that happens to be the case but that could have been otherwise: matter might have been distributed a bit differently at the birth of the solar system and clumped into a different number of bodies that satisfy the current definition of planet. What distinguishes a law is that, in some sense, things *must* satisfy it.

I say 'in some sense' because we can consider the possibility of different laws holding. In the next section we'll consider the Game of Life, in which the laws are very different from Newton's, but it makes sense to ask how things would be in that case. So what we mean is that out of all the things that are true, there are some that could have been otherwise without fundamentally changing the world, but there are some that couldn't. If there were a few more planets, then things might be very different, but if Newton's laws didn't work for the solar system, then the rules of the game would be completely changed. So the laws have to hold in a way that other truths don't.

How do physicists tell that something is a law rather than just true? Consider observations of the visible planets; we see that their motions obey Newton's laws. That is evidence that the laws are true, but how does it show that they *must* be? That's a pretty deep philosophical question, but we can say that as a matter of fact, the truths which are identified as laws are those that say the most about the world in the most compact manner. For instance, Newton's four laws (1–3 plus the law of gravity) form a very compact system, which allows us to calculate the motions of a huge array of bodies. That's why we think they are laws. But why should it follow that because a few words say a great deal they *must* be true? That's the philosophical puzzle.

To my mind it's just that being compact but descriptive makes the laws so incredibly useful and informative about the world that we think of them as the most important, most characteristic facts. And then of course it follows that we also think that the world would be different in the most important ways, that the world would have a completely different character if the laws were not true. That is, what we really mean by saying that the laws must hold is that otherwise the most powerful description of the world would be different. Many philosophers disagree with this answer, though fortunately our investigation will be independent of the debate, but I think I owed the reader some account to help explain the idea.

Next we need a clearer image of the range of worlds in which the laws are true: for instance, if Newton's laws held but there were fewer planets. So consider the following analogy; imagine a vast heap of books in which each book contains a different consistent story. Now imagine someone giving you a list of things all of which they wish to be true in a story, and your going through the books to pick out just those in which every item on the list is indeed true. Those books form a 'library' of all and only the books that are consistent with the list. The heap of books is in analogy to every logically possible history of the universe (including those with laws of physics quite different from ours); the list corresponds to the laws of physics; and the library you create corresponds to the physically possible histories of the universe (technically called the 'models' of the laws). Just as the list tells us what complete stories are allowed, the laws of physics determine which histories of the universe are possible.

For example, suppose the list is Newton's laws; each of the books would describe a possible universe in which they were true. Some would describe our planets orbiting our sun, others would have similar descriptions but involving fewer or more planets, and so on. But every one of the books describes a world that is possible according to Newton's laws. Similarly for any set of laws: the books picked out by the corresponding list differ in details, just as different things happen in the worlds in which the laws are true.

Before we see how this analogy can help us, let me make an important point about the use of the word 'theory'. When physicists (and philosophers and scientists generally) talk about a theory, they usually mean a collection of laws. What they definitely don't mean is that there is something tentative about the laws, as common usage has it. If they think the theory is tentative, then they will say so, describing it as, say, 'speculative'. 'Theory' in the sense of a collection of laws is the logician's and mathematician's use, and while scientific theories involve a bit more (for instance, associated examples, methods of calculation and experimentation, and so on), the sense is close enough.

Now we can say that theory is *future-history deterministic* if, according to the laws of the theory, the past history determines the future history. That is, in all the possible worlds that the laws of physics allow, you can't find two that are the same up to some time but differ thereafter—for any specific past history, the laws allow only one future history. In terms of our simile, any two books in the library corresponding to the theory that tell the same story up to a given point tell the same story from then on—they contain the same story. Some examples are Newton's laws of the motions of the planets, the laws governing a clock, the equation for the motion of a wave, the theory of an electronic amplifier. (Similarly, a theory is *past-history deterministic* if every pair of worlds that are the same after some time are the same before that time.)

Other theories are indeterministic, and more specifically *future-history probabilistic*: the past history and laws allow different possible futures but assign *probabilities* for them. For instance, imagine a world containing nothing but a coin, in which the laws are that the coin is tossed once a minute with equal chances of coming up heads or tails. Then there are possible worlds (or books in the corresponding library) in which the history goes *heads – heads – tails – heads – tails –*... but also worlds in which it goes *heads – heads – tails – heads – heads –*.... Such worlds agree for the first four tosses but differ after, so the histories do not determine the future. At any point, there is only the 50:50 probability of heads or tails the next time the coin is tossed.

The Game of Life

We sketched Newtonian mechanics, but it will be helpful also to work through a hypothetical physical theory in more detail. We'll make the theory as simple as possible while still being interesting. A really nice example is the 'Game of Life' invented by the mathematician John Conway in the 1970s (not the board game!).

Here's the theory. The world is two-dimensional and infinite and is divided into identical squares, or *cells*, which may or may not be inhabited by simple creatures at any time; if a cell is inhabited, we say it is *alive*, otherwise it's *dead*. The eight cells surrounding any cell are its '*neighborhood*'. See figure 1.4.

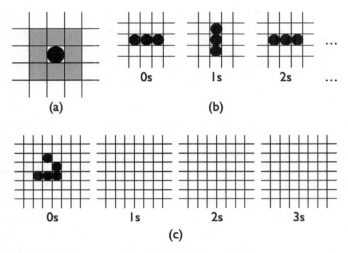

Figure 1.4 (a) The dots show which cells are currently 'alive'; the neighborhood of the live cell is shaded. (b) You can easily check that the laws of life allow this sequence of states. For instance, the leftmost living cell at 0 s has only one living neighbor, while the empty cell below the middle of the row has three. What comes next? (c) I've left the diagram blank after 0s so that you can fill it to make sure you understand the rules. (Hint: this shape is called a 'glider' in Life.)

Time, like space, is discrete, say with the smallest intervals lasting 1 s. Thus we can ask about the state of the world after 0 s, 1 s, 2 s, and so on. *State* is the term physicists use to mean 'how things are right now', a complete description of the physical properties of the world. For Newtonian mechanics, the state is a description of the locations, motions, and masses of all the bodies in the world, plus the forces on them. For Life the state is a complete description of which cells are alive.

The laws are very simple:

1. If a cell is alive at present and has exactly two or three living neighbors, then it stays alive in the next moment.
2. If a cell is dead at present and has exactly three living neighbors, then it comes alive in the next moment.
3. Otherwise a cell will be dead in the next interval.

Though the rules are very simple, some fascinating and complex histories are possible (see figure 1.4 again for simple examples): structures that move through space, systems that produce new matter continuously, and patterns that interact in astonishing ways. We don't have time to investigate these possibilities (though we will touch on some in later chapters); instead refer to Further Readings at the end of this chapter.

For now, the significance of Life is as a manageable example of a phys-ical theory; clearly with sufficient patience anyone could calculate what it predicts for succeeding times given an initial state. (All the talk about cells being 'alive' or 'dead' sounds biological, but all it really describes is whether there is something there or not.) Despite being incredibly simple, Life has some important features of typical physical theories, which make it a useful model. Principally it describes the evolution of things in space over time.

Unlike Newtonian mechanics, it does not describe how bodies move over time; instead it describes the properties of places in space over time. In fact many of our contemporary theories have this character; they are known as *field* theories. For instance, the theory of electromagnetic forces is a field theory; given the strength of electric and magnetic fields at different places at one time (and the locations of charges), the strengths at different places at later times are determined. Or again, general relativity (discussed in the next section) is a field theory, in which the field at any point describes how space is 'curved' there. The mathematics of Life is much simpler—you know the whole theory and can apply it yourself—the basic picture is a useful model of modern physics. We will use it several times in the book.

And of course the library simile works for Life too. Some sequences of alive and dead cells are consistent with the laws, but others are not; the former are the possible worlds of Life. Picture a library in which the books have a picture of the universe on each page showing alive and dead cells. Only books in which succeeding pages show allowed states are in the library; the books describing impossible Life worlds are expunged. (It should also be clear that Life is future history deterministic, since the state at any time determines for each cell whether it will be alive or dead next, thereby determining a unique subsequent state. Can you also see why it is not past-history deterministic? Consider what states might have preceded an entirely empty universe.)

1.4 SPACETIME TODAY

One more piece of background before we start: what does contemporary physics look like? This book doesn't aim to explain these theories in detail; we will see the important philosophical ramifications of rather simpler aspects of physics. Moreover, I want the reader to understand the physics that we discuss, rather than pull conclusions out of a hat. That said, recent developments undeniably have important consequences for some of our topics and it would be misleading not to explain how. So I'll give a sketch here and refer the reader to the Further Readings for more information.

First, Einstein's 1915 'General Theory of Relativity' or *general relativity*. The mathematics is the marriage of two ideas that we will investigate in later chapters: curved space (chapter 7) and relativity (chapter 14). But the basic idea is that space and time can be curved in much the same way that a rubber sheet can be laid flat or stretched and squashed into a wavy shape.

What makes them curve? The distribution of matter in space and time ('matter' here means energy in any of its many forms, including light). But what determines that distribution? The way space curves! In the pithy dictum of astrophysicist John Wheeler: "Matter tells space how to curve, and space tells matter how to move." That is, a world is possible according to general relativity (i.e., corresponds to a book in the 'library' of general relativity) if matter and space are mutually related in the correct way.

There are many important consequences of this theory. First, it gives a new understanding of gravity, not as a force but as motion along the 'straightest' lines in space and time. The Earth does follow the straightest path around the sun after all, but the mass of the sun curves space and time in such a way that that path is an ellipse in space! Second, in the 1960s physicists Stephen Hawking and Roger Penrose discovered that possible worlds that look like ours each began as a single point a finite time ago—the 'big bang'. That is, look in the library at any of the books on general relativity; if the book describes a universe like ours, then it has a beginning and describes a universe exploding from a point. Finally, the theory also predicts that there can be so much matter in a region that space around it curves so much that matter, even light, cannot escape at all. Such a region is a 'black hole'.

But general relativity doesn't seem like the whole story. Whenever we look at very, very small things—atoms and the like—we find that they are described by a theory of *quantum mechanics*. But general relativity is not a quantum mechanical theory, so physicists don't have a theory of gravity for very small things. To attempt to explain quantum mechanics here would be too much effort for too little reward in this book. (We will consider quantum theory in chapters 17 and 18, but only one of its more easily understood aspects.)

What we do need to know is that the most popular, most developed quantum account of gravity is *string theory* (see the Further Readings). This theory really is speculative, in part because it is so mathematically complicated, but many extremely bright physicists think that it is worth pursuing (and they are gambling their careers, of course). The picture of string theory is that the fundamental constituents of the universe are tiny one-dimensional 'strings', some open ended and some looped, moving in a space of many dimensions, itself wrapped in an ornate pattern. Gravity and all the particles and forces of nature are to be understood as different vibrations of these strings, much as different notes on a guitar are understood in terms of different vibrations in the strings.

Obviously there's quite a story about why a world like that would look like ours, and especially how that explains gravity; but again, it is nicely explained elsewhere. (In chapter 6 we will discuss why some dimensions might be unobservable.) What will suffice for our purposes is the basic picture: tiny vibrating strings, moving about in a complicated space.

Just how tiny are these strings? Well, to give a rather arbitrary comparison, the kinds of distance at which the world would simply look like strings is to a proton as a proton is to the city of Chicago! That is, very tiny indeed!

We now have all the ideas of physics and philosophy that we need to get started. So let's go.

Further Readings

The passage from Melissus is found in Richard McKirahan, *Philosophy before Socrates* (Hackett, 1994, 293). The quotation from Descartes comes from part II (sections 23 and 64) of his *Principles of Philosophy* (Reidel, 1983), translated by Valentine Rodger Miller and Reese P. Miller.

The best source on Life remains Martin Gardner's columns in *Scientific American* (reprinted as chapters 20–22 of *Wheels, Life and Other Mathematical Amusements* [W. H. Freeman, 1983]).

There's a nice java implementation at http://www.bitstorm.org/gameoflife/.

Brian Greene's *The Fabric of the Cosmos* (Knopf, 2004) explains general relativity, quantum mechanics, and string theory accurately and clearly at a level suitable for a general audience. I recommend it strongly. For a little more rigorous introduction to general relativity I suggest Robert Wald's *General Relativity from A to B* (University of Chicago Press, 1978); it explains the fundamental principles beautifully using nothing but some simple algebra and geometric intuitions. Finally, for an insight into alternatives to string theory, I recommend Lee Smolin's *Three Roads to Quantum Gravity* (Perseus Books, 2002).

2

Zeno's Paradoxes

Having settled to some extent the way in which we should conceive physical change, we turn now to some old arguments, attributed to Zeno, against the mathematical description of change. First, what do I mean by 'the mathematical description'? Just the ideas that (1) some properties—distance traveled, time taken, weight, wealth—can be given numerical values and that (2) we can find interesting mathematical relationships between them, such as that the taller a professional basketball player is, the wealthier she is likely to be. These are of course very natural assumptions for properties such as distance and weight, but neither assumption is trivial. We don't really think that happiness is quantifiable, for instance, though it certainly comes in greater and lesser amounts, and there is (I imagine) no mathematical relationship between local time of birth and adult height. That said, the idea that many processes do have mathematical descriptions is a central assumption of the mechanical and Newtonian conceptions of change, and it was certainly believed by Aristotle. Thus Zeno's arguments are attacks on change viewed in any of the ways we have considered.

We don't know much about the life of Zeno except that he was born about 490 B.C. and (according to Plato's gossip) that in addition to being Parmenides' pupil he was also his lover, not an unusual occurrence in academic life during any period. According to Plato, Zeno argued that those who ridiculed his teacher's view that there is no change were in fact logically committed to views equally absurd. Specifically, although they believed that change occurs, they had no consistent (mathematical) description of it. I would say that in terms of the mathematics of his day, Zeno was correct, but that modern mathematics (of the nineteenth century and beyond) does allow a consistent treatment. To see these points we'll look at a couple of his paradoxes, the Dichotomy and (in the next chapter) the Arrow. We will see that even our 'good' contemporary solution to them reveals some surprising features of change. Along the way, we'll consider some other related proposals that Zeno's later admirers have devised.

2.1 THE DICHOTOMY PARADOX

As the name suggests, this paradox arises when we consider division into two, without end. For instance, imagine Atalanta (the mythological athlete and hunter) sprinting down a 100-meter track. Think of her run in the following way: first she must run the first 50 m, then she must run the next 25 m, and then the following 12.5 m, and then 6.25 m and then. . . . We can play this game endlessly (since half of every finite remaining length is itself finite) and so even a run with no stops can be thought of as comprising an infinite number of smaller runs. Zeno then pricks our intuitions: how can an infinite number of finite runs, however small, ever be completed? (Indeed, Zeno took it to be 'self-evident' that an infinity of tasks can't be completed.) And if not, then Atalanta cannot reach the end of the track. Worse, since the argument can be repeated for any section of the track, she cannot reach anywhere at all! She is (and by analogous argument, all things are), it seems, immobilized, simply by mathematics. (This line of thought can be applied to many cases: a friend once commented on the 'Zeno's papadom' effect, often experienced in Indian restaurants, in which diners take successive halves of whatever remains of the final papadom, making it an infinite meal.)

The natural reaction is to point out that we clearly have lots of experience of things moving and dismiss the conclusion as absurd; when Diogenes the Cynic heard Zeno, he sought to disprove him by wordlessly standing up and walking out. I, like most people, agree that the conclusion is absurd, but simply to dismiss Zeno on that account is to misunderstand the logic of this kind of paradox.

We are faced with three choices: (1) Accept the argument and its absurd conclusion. We could then account for our experiences (as Zeno probably would have) by putting them down to some kind of systematic hallucination, so that Diogenes merely *seemed* to be moving—a triumph of reason over experience. This choice is not very palatable, except to extreme paradoxers.

(2) We could (as Diogenes seems to have done) deny the conclusion without demonstrating a flaw in Zeno's argument. People often do reject conclusions that they don't like without explaining what is wrong with the arguments that produce them—you can see it happening any Sunday morning on the political talk shows—but doing so is not thoroughly honest. On the face of it, Zeno makes only true assumptions and takes only logical steps, so logically whatever he infers from these must be just as true. Hence, if you want to get off the hook, you need to explain what went wrong: either what assumptions are false after all, or how the steps are not all logical. Thus although 'refuting' an argument often means just denying its conclusion, such refutation should show why the argument doesn't work.

(3) The third possibility then is to show what (if anything) is wrong with Zeno's argument. In so doing we not only live up to the standards

of rational integrity, but also receive the benefit of refuting an argument, namely learning something new: what was wrong with those assumptions or inferences that seemed right to us to begin with.

Aristotle put his finger on the problem: suppose for convenience that we believe that Atalanta takes 10 seconds to complete her sprint at a constant rate. Then we can divide up the interval of time in just the same way that we divided up the distance, so she has 5 s to complete the 50 m part, 2.5 s for the 25 m part, 1.25 s for the 12.5 m, and so on; for every one of the infinite number of runs that make up Atalanta's trip there is an interval of time within the 10 s. In other words, since time and distance can be divided in the same way, then an infinite series of tasks— this one at least—can be completed in a finite time, contrary to Zeno's reasonable sounding assumption, which now is seen not to be reasonable at all.

Aristotle may have addressed Zeno's argument, but, as he recognized, his refutation raises a new problem. We've divided 10 s into an infinite number of finite parts, but shouldn't an infinite number of finite anythings add up to infinity? After all, every new addition makes the total larger. And so doesn't it follow that 10 s is actually an infinite time (or, put less paradoxically, that there can be no finite times)? If so, then Atalanta still takes an infinite time to reach the end of the track, or to cover any other distance, which is to say that she never does.

Aristotle's response was to say that if we 'actually' had an infinite number of intervals, generated by, say, snapping one's fingers at the beginning of each, then they would indeed add up to an infinite time. However, no finger snapping or anything similar does occur, only a division 'in thought', so we have only a 'potential' infinity of intervals; we could but don't divide up the 10 s. According to Aristotle, while an actual infinity of intervals would sum to infinity, a merely potential infinity does not: somehow there's nothing actually to add together.

This answer, though, is far from satisfactory: in the first place, surely 10 s has all those parts whether or not we do something concrete to split them up. (Isn't 10 s actually comprised of two 5 s intervals whether or not we snap our fingers halfway through? And if so, then isn't it also composed of the infinity of halves whether or not they are divided by finger snappings?)

In the second place, even if an interval is actually divided into an infinity of finite intervals (each separated by a finite time from the next), then it can still be finitely long. (We shall discuss how when we turn to 'supertasks' shortly.) So Aristotle is just wrong to think that actual infinities of finite quantities must be infinitely big.

Finally, what is at stake is the possibility of a *mathematical* description of space and time. But the *only* way that we can divide mathematical objects is in thought (to be fair, Aristotle might not have agreed with this point). So a mathematical line of length 10 units is comprised of an actual infinity of parts and hence seems not finite after all; therefore it can hardly represent a finite interval of time. We need a better response than the theory of actual versus potential infinities.

What we need to say instead is that $10\,s \times (1/2 + 1/4 + 1/8 + \ldots) = 10\,s$. Given that we generated the series of intervals by dividing $10\,s$, this equality seems intuitively plausible; however, care is needed. Consider for instance the infinite sum $x = 1 - 1 + 1 - 1 + 1 - 1 \ldots$ What does this equal? We could rewrite it as $x = (1 - 1) + (1 - 1) + (1 - 1) + \ldots = 0$. Or we could write $x = 1 - (1 - 1 + 1 - 1 \ldots) = 1 - ((1 - 1) + (1 - 1) + (1 - 1) + \ldots) = 1 - 0 = 1$ (using our first result). But should we say now that we have two equally good calculations, one showing that $x = 0$ and one that $x = 1$? Since $1 \neq 0$, we must conclude that the calculations are in fact equally *bad*.

The problem is that we tried to apply reasoning familiar from finite sums to an infinite sum and thereby were led astray. For instance, if we have to add an equal *finite* number of alternating $+1$ and -1 terms, then we group them together to form a finite sum of 0 terms: for example $1 - 1 + 1 - 1 = (1 - 1) + (1 - 1) = 0 + 0 = 0$. It's intuitive to suppose that the same reasoning can be unproblematically applied to infinite sums, but we have just seen that carrying over the rules of finite sums to the infinite case leads to nonsense. But if that intuition fails, then we should also question our intuition that $1/2 + 1/4 + 1/8 + \ldots = 1$ and look for a careful understanding of infinite sums.

We can find it in a key distinction among infinite sums: $1/2 + 1/4 + \ldots$ gets ever closer to, but never exceeds 1 (and exceeds all numbers less than 1 at some stage); $1 - 1 + \ldots$ alternates between 1 and 0; and, for instance, $1 + 1 + 1 + \ldots$ eventually is larger than any finite number. Technically, 1 is the 'limit' of Aristotle's sum, while the other two have no finite limit. We can say then that an infinite sum is equal to its limit, if it has one. If it does not have a limit, then we shall say that it is *undefined*, like $1 - 1 + \ldots$, reasonably, given the ambiguity that we discovered. And if we say this, then we do indeed find that $10\,s \times (1/2 + 1/4 + 1/8 + \ldots) = 10\,s \times 1 = 10\,s$. What is interesting is that this approach was not fully given until the nineteenth century, by Augustin-Louis Cauchy, whose name was given to such sums (if you have ever studied elementary calculus you will surely have come across them before).

But why do we only 'say that' this is how infinite sums are to be understood? Haven't we *discovered* what infinite sums are? No, if we think carefully about the relationship between familiar finite sums and these infinite ones, what we have done is give a new rule of addition. For, as the example shows, the rules of finite addition do not determine the result of infinite additions; they make 0 and 1 equally good. Thus we need to think of the 'Cauchy sum' as a new mathematical function, though one defined in a reasonable way in terms of our existing finite sums. In that case one might worry that we have not really addressed Zeno: Aristotle's first response just asserted that the infinite sum of intervals was $10\,s$, and haven't we just done the same? No, we have done more: we have given a precise mathematical rule that applies to any infinite sum (even those it makes undefined), whereas Aristotle gave no such rule. We have *shown* that mathematics can accommodate a suitable notion of infinite sums.

One might ask finally how we can show that this conception of infinities is the right one for describing nature; why not the 'bad' rule that all infinite sums equal infinity, for instance? But here what is needed is an empirical argument like that of Diogenes, showing that Cauchy's conception matches our experiences. (Things did not have to be that way; one could imagine time being finite and composed of finite, indivisible 'atoms', so experience is relevant.) Continuing the argument further would require showing somehow that despite the viability of a mathematical description of change, there is some other reason to doubt our experiences. But what?

2.2 'SUPERTASKS'

Various philosophers have advocated a modified version of Zeno's position. 'Sure,' they say, 'mathematically, Atalanta can reach the end of the track, but that is so because she in fact has only one run to complete, and conceptually dividing it into half-runs doesn't alter that fact. But it is impossible for her to complete an infinite series of *separate* runs in a finite time. Indeed, as Zeno claimed, it is impossible for an infinite number of *distinct* tasks—collectively, a "supertask"—of any kind to be completed.'

They might even go further and argue that if change of any kind were possible—including Atalanta's run—then supertasks would be possible; and if supertasks are not, then neither is any change. (These philosophers are well aware of Cauchy's rule for infinite addition, so what they have in mind is that it does not apply to any infinity of distinct actions; nor, if they carry things further, to any infinity of tasks, distinct or not.)

There are myriad examples offered to back up the claim that supertasks cannot be completed, but we'll just look at—and debunk—three especially startling ones. First, it turns out that Atalanta could complete her run even if it were divided into an infinite number of runs, each separate from the next by a period of rest: that she could finish even if her run were a supertask! (Put another way, as we noted above, she can complete an 'actual' infinity of runs.) Drawing on an idea of philosopher Adolf Grünbaum's, here's how she could do it: each leg of her 'staccato' run involves running 3/4 of the remaining distance, at 1/2 the speed of the previous leg, then resting for the same time as she ran. Thus, for instance, she might run the initial 75 m at 30 m/s, taking 2.5 s; then she rests for 2.5 s; then she runs the next 18.75 m at 15 m/s, taking 1.25 s; then she rests for 1.25 s; and so on. When she has completed 100 m she stops. Her run is shown in figure 2.1.

In this sequence of runs, each leg is always 1/4 as long as the previous one (e.g., 18.75 m is 1/4 of 75 m), so the total distance covered

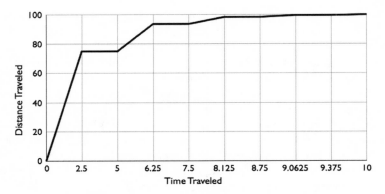

Figure 2.1 The Staccato Run. Atalanta's position is plotted against the time, showing her running at decreasing speeds over shorter times, with rests in between.

is 75 m × (1 + 1/4 + [1/4 × 1/4] + [1/4 × 1/4 × 1/4] + ...), where the terms represent the lengths of each leg. According to Cauchy's rule (if you work it out), this sum is just 75 m × 4/3 = 100 m, so Atalanta's run definitely is the full length of the track. Of course we also need to check that she finishes in a finite time.

Well, each run is 1/4 the length of the previous one, but it is traversed at only 1/2 the previous speed, so it takes 1/2 the previous time (e.g., 1.25 s is 1/2 of 2.5 s). So the total time she takes is 2 × 2.5 s × (1 + 1/2 + 1/4 + ...), where we double the total because she always rests a time equal to the run. The sum is just 1 plus Zeno's original sum, so the time she takes is 2 × 2.5 s × (1 + 1) = 10 s again. That is, after 10 s, she has traveled 100m and then stops, the infinite series of distinct runs completed. (In Grünbaum's version of the staccato run, Atalanta runs every interval at the same speed, which, he shows, implies that she requires an infinite amount of energy to complete it. In my run Atalanta cuts her speed in half for each run, and she only needs a finite amount of energy.)

So the staccato run does not provide support for the Zenoan opponent of supertasks, since it is a supertask that can be completed. Let us turn to the second example, known as 'Thompson's lamp', and see whether the same can be said of it. Imagine a table lamp with an on–off switch: at 1 minute before noon the switch is pushed 'on', then at 30 seconds to noon it is turned off, then at 15 s to on again, then at 7.5 s to off, and so on, each subsequent switch occurring after half the time of the previous one.

We'll imagine that the effect of the switch is instantaneous (actually we'd also better make it an unmoving, touch-sensitive switch and specify that on each depression the finger pressing it moves half the previous distance, so that neither switch nor finger moves an infinite distance during the series). We know then that there will be an infinite series of switchings in the minute leading up to noon, since the series of times between switchings have the sum $60\,\text{s} \times (1/2 + 1/4 + \ldots)$, which we know from Atalanta's original run equals (by the Cauchy rule) 60 s.

The puzzling question is, 'Is the lamp on or off at noon?' After all, the supertask is finished at noon. It seems that the lamp can be neither. The infinite sequence of switchings cannot finish with the lamp on, because for every step for which the lamp is on, there is a step right after for which it is off; similarly, the sequence can't end with the lamp off, because whenever the lamp is off in the sequence there is a time just after at which it is on. Since, it seems, the lamp would not end up in either of its two possible settings—on and off—Thompson's supertask appears impossible. But why should the supertask be impossible? Why else but because the individual tasks are impossible? That is, shouldn't we conclude that switching a light on or off is impossible? And if such a simple change is impossible, surely Zeno was right that any change is impossible!

Again, there is a fallacy in this reasoning. Thompson's description of the supertask specifies what happens to the lamp starting at 11.59 A.M. but *before* 12 P.M.—at any such time we could work out whether the lamp was on or off—while it simply doesn't say anything at all about noon or after. Put another way, the sequence is over at noon. So one cannot, after all, argue that the lamp must not be on (or off) at noon because whenever it is on (or off) it is off (or on) just after; that reasoning applies only before noon. Thus Thompson's description tells us as much about the lamp at noon as the knowledge that there is a woman in the next room tells us about the color of her hair. Without more information, the lamp can be either on or off, just as her hair can be either blonde, brunette, or whatever.

This discussion points to one remarkable feature of the supertask: while there is a time when it is over (noon and after), it contains no last task; every single switching has another switching after it! Indeed, one way to understand why Thompson's story doesn't determine the state at noon is to see that the endless series of switchings, finished before 12:00, has no last element and so fails to leave the lamp in any *final* state that it must possess at noon. So Thompson's lamp points to something really quite counterintuitive, but quite logically possible. Since any finite interval of time can be divided into an infinity of finite parts, time is structured to allow infinite series of tasks to take finite times—time permits tasks to be completed even though they have no end! (Of course, both of Atalanta's runs are also like this: even when she runs without stopping, there is no *last* half of the remaining track that she covers, and yet there she is at the end after 10s!)

Note also that there is nothing unusual about having too little information to make a prediction. Suppose I tell you that two cars are approaching each other at a constant 10 m/s and ask when they will collide. You can't tell unless you also know their positions. What is a little unusual about Thompson's lamp is that one can't predict whether the lamp will be on at noon *even though one apparently knows everything about the state of the lamp at 11:59 A.M.* Contrast this situation with that of the cars: I made the prediction impossible in that case by withholding information about the states of the cars (their separation); once that information is known, the prediction can be made, because the laws are deterministic. However, the 'law' governing the lamp—that it is switched after half the previous time—is indeterministic, since the state at 11.59 does not determine the state at 12 P.M. (or even probabilities for the state).

Now if a real Thompson lamp were constructed, in accordance with real deterministic laws of physics, then its state and the state of the switcher at 11.59 A.M. would have to determine the state at 12 P.M., by definition of 'determinism'. That is, in any realistic setup there must be some aspect of the lamp or switcher's state that is is not mentioned in Thompson's story but that determines, according to the laws of physics, the state at noon. If we knew not only that it was on, but also perhaps how its molecules were arranged, we could in principle determine its state at 12 P.M. Thompson's story requires either that the laws of physics are not deterministic (indeed, not even probabilistic, since the story doesn't yield probabilities for the lamp at noon) or that an incomplete description of the setup is given.

A third supertask paradox ('Bernadete's') is interesting because the 'paradox' does not even require that the task actually be undertaken, just that beings be prepared to undertake it if necessary. Suppose that the gods have decided that they do not want Atalanta to reach the end of the track, but that each plans to prevent her in a different way. The first decides to prevent her from going beyond halfway, by placing a wall immediately in front of her, if she reaches 50 m. The second decides not even to let her go beyond a quarter of the way, by placing a wall at 25 m, if she reaches there. The next will similarly place a wall at 12.5 m, if she gets that far, and so on and on, each of the infinity of gods deciding to stop her at half the distance of the next god, but only if she gets that far. (In fact, since placing walls does not require divine intervention, and since no one is required to have infinitely fast reactions, there is actually no need to use gods; an infinity of mortals should be capable of the task.)

Now suppose that nothing else is hindering Atalanta, and that she is sufficiently determined and able, so that she will run unless prevented by one of the gods' walls. Then we have a contradiction, for she in fact cannot get anywhere, even though none of the walls is raised! Pick any distance along the track, say 10 m. She can't have gotten that far, because she would have been prevented by a wall at 6.25 m. What about 5 m? No, the wall at 3.125 m would have stopped her. And so on—pick any distance, there's a *potential* wall before it, which will stop her getting that far if she runs. But if she doesn't actually get anywhere, then none of the gods will make those potential walls *actual*; there will be no walls.

But then . . . nothing stops her running, so she does (which of course she can't)!

On the face of it, it doesn't seem that there is anything impossible with the task given to the gods. It seems that the only way to make the story consistent is to say that despite her intentions, and despite the absence of any physical barrier, Atalanta is not after all capable of motion. And if we accept that conclusion in this case, we can reflect that the gods don't actually *do anything*, and so they surely aren't what prevents the run; it must be that running itself is impossible and we are back to Zeno's conclusion about motion in general.

However, it is not Atalanta's motion that is impossible in this situation, it is the gods' plans—their 'superintention'. Surprisingly, the whole infinite pantheon of gods cannot satisfy their collective intentions—a surprising fact, because each god seems capable of fulfilling his particular intention to build a wall if Atalanta reaches a certain point.

To see why they cannot, first consider a simpler example of how individuals' intentions can interfere to prevent them from being carried out, even though each seems possible on its own. Suppose Arlo loves Zowie but hates parties, while Zowie loves parties but can't stand Arlo. And suppose there's nothing psychologically or physically preventing either Arlo or Zowie from going to some party. But it is still impossible for Arlo to intend to go to the party if Zowie does, and not otherwise; for Zowie to intend to stay away from the party if Arlo goes, and go herself otherwise; *and* for both of them to satisfy their intentions. If both go, Zowie fails to avoid Arlo; if neither goes she misses the party she wanted to go to; while if just one goes, Arlo either misses the chance to see Zowie or goes to a party he didn't want to. But those are all the possibilities, so it is impossible for them both to realize their intentions.

The conflict, of course, arises because the intentions of each are conditional on the actions of the other, which are in turn dependent on the intentions of the other. Thus their plans are interrelated and get in each other's way, even though both are in principle capable of going to or not going to a party.

Much the same thing happens with the gods; each is capable of building a wall, but each has the intention of building a wall if and only if none of the previous gods do. If none builds a wall, then they all fail; if there is a first one to do so, then all the earlier gods fail; and if they all do, then each builds a wall despite the presence of earlier walls—and they all fail again. But those are the only possibilities, so like Arlo and Zowie, the gods' plans are incompatible!

So Atalanta's motion is saved again: the argument can be seen to be saying that *if* the superintention can be carried out, *then* Atalanta cannot move. But that would be like imagining a situation in which I intend to make $2 + 2 = 5$ if Atalanta moves, and concluding that she cannot move. In either case what is impossible is not her motion, but fulfilling the intention.

In this chapter we've seen a number of ways in which mathematical physics responds to the challenges posed by Zeno's paradoxes and their cousins. Clearly there's some kind of interaction going on here between physics and philosophy: physics has a picture of motion, and philosophy questions its cogency. However, we will postpone until the end of the next chapter the question of exactly how to think about the interaction. The chapter will first investigate another kind of paradox given by Zeno.

Further Readings

You can find an online article that I wrote on Zeno's paradoxes at http://plato.stanford.edu. Plato briefly discusses Zeno in his *Parmenides*, while Aristotle's responses come from his *Physics*, Book VI.

My discussion of Bernadete's supertask follows a paper by Stephen Yablo in *Analysis* (2000: 148–151).

3

Zeno's Arrow Paradox

In the previous chapter we considered real and alleged paradoxes concerned with breaking a finite task—a run, for instance—into an infinity of smaller tasks. We've learned the mathematics created to deal with such issues, the mathematics that lies at the foundation of physics, which permits a mathematical treatment of change. In this chapter we turn to another kind of paradox, related to the way we understand time: the paradox of the arrow. Once we have understood this argument, and what it shows, we will discuss what we can learn from Zeno about the ways in which philosophy can teach something to physics.

3.1 THE PARADOX

Zeno's second line of attack on motion makes use of the idea that time has smallest parts. Intuitively we think of these parts of time as lasting exactly zero seconds—an 'instant'—but Zeno's argument would work just as well (or badly) even if the smallest parts of time were finitely long. This point is worth noting because according to quantum theories of space and time, such as string theory, there likely are smallest parts of time, called 'chronons', which last for 10^{-44} s—the 'Planck time'. The idea is that any change whatsoever will take a whole number multiple of this 'quantum' of time; so the quickest change still takes 10^{-44} s. Clearly in this case the assumptions of the dichotomy fail; the times taken for each part of the run cannot be half the previous time without limit, for no part of the run can take less than 10^{-44} s, and the sum of times has a finite number of finite terms.

However, Zeno's 'Arrow Paradox' works the same whether the smallest parts of time are unextended instants or finite chronons. We will assume they are instants, but only for simplicity. The argument goes like this: consider an object in motion, traditionally an arrow, and break its journey down into the smallest possible parts. By definition, the smallest parts don't have earlier and later parts, since they would be even smaller. And so the arrow does not move during any instant; if it did move, then we could distinguish earlier and later parts of the instant as 'the time when the arrow was there' versus 'the time when it was a bit further forward'.

(Of course this conclusion is also in agreement with our idea that an instant has no duration: the arrow has no time to move during an instant of zero duration.)

But what is any interval of time but a collection of instants (presumably infinitely many)? Between any two instants are only other instants, not any 'temporal glue' holding them together. "So," Zeno appears to argue, "when does the arrow move? If there's nothing to the interval but instants, then the only times that the arrow could move is during each of the instants; but the arrow cannot move during any of them, because they are indivisible by definition. Thus it has no time at all for motion, and so cannot move!"

There's something quite wrong in this line of thought, but also something quite important. In the first place, it commits the 'fallacy of composition': just because everything in a collection has some property, it does not follow that the collection has that property. Everyone in the United States weighs less than a metric ton, but the total weight of the population is greater. No atom of my brain can think, but taken together they can. And just because no instant of time is long enough to allow motion, it doesn't follow that an extended interval doesn't allow it—though we do owe an explanation of how that is consistent with immobility during any instant. (Just as we owe an account of how my brain can think.)

The idea is simply that motion amounts to being in a continuous (meaning no jumps, in a precise sense) series of positions over the continuous series of instants that make up an interval of time; *at* each instant the arrow is simply *at* an appropriate place. For instance, if the arrow is moving from 0m at noon at a rate of 10 m/s, then at t seconds after noon, it will be $10 \times t$ m further on—to each instant a position. This theory is usually known as the 'at–at' theory of motion, for obvious reasons. It is in fact a description in words of the modern mathematical account, which, like Cauchy's sum, was fully developed and understood only in the nineteenth century.

The at–at account might seem like a cheat, but I'd like to show that it isn't, through a series of objections and responses.

The general question is 'How does the arrow get from one place on the trajectory to another, if it never moves during any instant?' The at–at answer is simply that motion involves nothing more than being at the right places at the right times. Any further requirement of getting from one place to another is rejected. (If time is, as we intuitively picture it, a series of zero duration instants arranged like points on a line, then there is something else to say. For there is no 'next' point on a line; between any two is a third. Thus in the intuitive picture, the question makes no sense.)

Then a more sophisticated question, 'If the arrow doesn't travel any distance during an interval, its speed—distance traveled per second—during any instant must be zero, but if its speed is always zero, how can it move?' If we suppose, intuitively, that instants have no duration, then the conclusion doesn't follow: zero distance divided by zero time does not make mathematical sense, and so the immobility of an arrow during an instant says nothing about the speed of the arrow. Instead, according to the at–at theory, the speed of an arrow at a particular time depends on its motion over a finite interval around the time. That is, the average speed over any interval is the distance traveled divided by the length of time. To get the speed at a time, we consider the average speed over smaller and smaller intervals; the speed at the time is the average speed as the interval shrinks to nothing. (If instants do have duration, then we have to accept a new definition of motion in which a finite speed over more than one instant is compatible with zero speed over an instant.)

What follows is that the states of an arrow at rest and of an arrow moving with any speed whatsoever, insofar as they pertain strictly to *a single moment*, are identical; what distinguishes them is the different places they occupy earlier and later. And so, while we can define an instantaneous velocity, there is a literal (though unconventional) sense in which there is no such thing. This point is the important lesson of the arrow paradox.

3.2 WHAT PHILOSOPHY CAN TEACH PHYSICS

Because of their incredible conclusion, Zeno's arguments are often dismissed without much thought. I hope our discussion has convinced you that that would be a mistake; we have taken the arguments seriously and found that they pose serious questions and problems for our intuitive conceptions of space, time, and motion. The answers and solutions that we have given required both major advances in mathematics and clarification of the relation of mathematics to space, time, and motion. The journey has led us to considerable insights, which we might well have missed if we had not been started on our way by Zeno's strange-sounding arguments. In this way, Zeno's philosophical analysis of motion, and what followed, produced important advances in the most basic concepts of physics.

Bertrand Russell summed things up this way:

> Having invented four arguments all immeasurably subtle and profound, the grossness of subsequent philosophers pronounced him to be a mere ingenious juggler, and his arguments to be one and all sophisms. After two thousand years of continual refutation, these sophisms were reinstated, and made the foundation of a mathematical renaissance.

Russell may be exaggerating Zeno's influence (and Aristotle, as we saw, certainly took his arguments seriously), but the point is that mathematics didn't receive the foundation demanded by the paradoxes until the nineteenth century and the work of people such as Cauchy. Before moving on, let's reflect on how Zeno's philosophical arguments had an effect on math and physics; how did he make philosophy relevant to physics?

In the dichotomy Zeno looked at the concept of space as described by geometry and exposed a gaping hole in the understanding of how the length of a line might relate to the lengths of its parts when there are infinitely many of them. On the one hand there was divisibility, while on the other an incomplete understanding of adding lengths. In the arrow, Zeno gave a similar critique of the concept of motion, which relates space to time: motion is the change of place with respect to time. Here he revealed a gap in the understanding of how motion with respect to the smallest parts of an interval relate to motion with respect to the whole. How is the absence of motion during any instant compatible with motion over an infinite number of them? Both these problems were solved by reformations in the foundations of mathematics, which provided clear, precise accounts of the problematic concepts: the Cauchy sum and the at–at theory.

Such careful exploration of fundamental concepts, the teasing out of contradictions and confusions, and the development of clearer concepts are a hallmark of philosophy. It is what Plato did in his dialogues for more familiar notions such as 'good' and 'knowledge'; Zeno simply started the same process for technical notions of space, time, and motion. One reason that philosophy of this kind can 'teach' anything to a technical subject such as physics is that even mathematical concepts can fail to be sufficiently well defined to apply to every kind of case that they are supposed to: How are notions of length applied to an infinity of distances? How does motion apply to the smallest parts of time?

Physics, of course, proceeds by experiment and by insights into how things interact, but as we shall see, it also advances by analysis and revision of fundamental ideas—by philosophical investigation. (Of course, while such work is philosophical, it need not, as we shall see, be carried out by people we would identify as philosophers rather than by scientists.) We shall see more examples of this kind of positive interaction in subsequent chapters, but we shall also see the influence running in the other direction, in which advances in physics force us to change our philosophical views.

Further Readings

A number of the works referred to in the last two chapters can be found collected in Wesley Salmon's (1970) *Zeno's Paradoxes* (reprinted by Hackett, 2001). His introduction and a number of the papers included would be suitable for those seeking to develop what they learned in this chapter.

Russell's verdict comes from his *Principles of Mathematics* (Cambridge University Press, 1903: Section 325).

4

The Shape of Space I

Topology

The discussion in chapter 1 led us to the idea that physical changes are in large part (though not entirely) spatial changes. But if space is where change happens, what kind of thing is space? Well, as a first description, it is the collection of all places where things either are or could be—the collection of all the possible places. But what more can be said? In the next six chapters we will investigate three aspects of space. What is its shape? Is it flat or curved? And what is the relationship between space and the material objects that inhabit it? These questions will let us get a handle on the properties of space, and on the issue of what kind of thing it is according to physics.

Topology is the branch of mathematics concerned with the possible shapes of space in a special sense (mathematicians are usually thinking about abstract mathematical spaces, not physical space). Imagine space as a stretchy rubber sheet; its *topological* properties are those that don't change however much it is stretched or squashed (without making cuts or joins). So we see in figure 4.1 that if the sheet is initially flat, one could push down in the center to make a 'well' without changing the 'shape' in the sense we are interested in. Or it could be squashed into the shape of a tall (handleless) beer glass. Or be given a wavy surface. Or be twisted into the shape of a riding saddle. All these surfaces have the same shape in the topological sense. The sheet, however, cannot be turned into a sphere without joining its edges, so it does not have the same shape; but a sphere can be squeezed into the shape of a peanut shell, so that has the same shape as the sphere.

Here, for brevity, we're using 'shape' to mean 'topological shape'. According to common usage of the word, in contrast, flat sheets and saddles, for instance, do not have the same shape; shape in this sense is studied in geometry (discussed in the next chapter).

Saying that topological properties are those that a rubber sheet will keep when stretched is clearly only to give a rough definition, useful for a heuristic idea of the precise mathematical notion of topology. Still, there are two topological properties that have particularly bothered those trying to understand space, which the rubber sheet analogy illustrates fairly well.

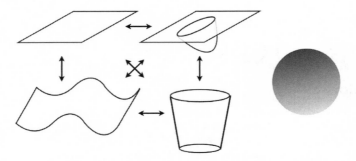

Figure 4.1 A flat sheet, a well, a beer glass, and a wavy sheet all have the same topology—they can be stretched and squashed into one another—but they do not have the same topology as a sphere.

First, if a sheet has a definite edge, then it cannot be removed by stretching: whether space has an edge or not is a topological property. Second, if a sheet is in fact perfectly thin, then it cannot be stretched into a block with some thickness (and conversely, a three-dimensional solid sheet cannot be squashed perfectly flat): the number of dimensions is a topological property. This chapter is concerned with a discussion of these two properties of shape.

4.1 AN END TO SPACE?

Most of the time, when we think about the space we live in, if we think about it at all, we think in terms of some surrounding locality. How local depends on the context: the placement of furniture in a room, the geography of our neighborhood if we want to walk to the store, the geopolitical divisions of the world, the configuration of the solar system perhaps, maybe even the shape of the Milky Way (a spiral) and beyond. Each enlargement of scope requires an abstraction from the sense of our immediate place, and seeing it as just a part of something larger.

The end point of this process of abstraction is to conceive of the space of *everything*. For me, and likely many others, the first conscious consideration of the whole of reality—the universe—was a remarkable and memorable moment in my intellectual biography. It was in Sunday school, and we were learning about the idea of unlimited knowledge, power, and goodness, so it was natural that the topic of the universe came up, since that is the 'all' to which these things apply. But as soon as you have that idea, and try to give it a concrete formulation, the question of its shape is unavoidable, especially the issue of whether it goes on forever, or has an edge (not, as we'll see, the only possibilities). We can't know whether the question arose at the same time that humans began to conceive of the universe itself, but, as we shall see, the question of edges is an old one.

The connection of the question to God's supposed unlimited attributes is strong: for instance, both Descartes and Newton addressed the question of whether anything beside God—in particular, space—could have infinite properties. But the question also bears on philosophical inquiry into the concept of existence—'ontology'—because of the idea, summed up in Aristotle's dictum, that 'whatever exists, exists somewhere'. If that is so, the limits of space are quite literally the limits of everything. We will bring such considerations to bear on our question, but we also want to see what physics can teach us about it.

Let's start first with a concrete model of a bounded universe and then consider an objection to the model. The model is Aristotle's, a sphere, including all its contents, or 'closed ball'—'closed' to indicate that the surface is included, and 'ball' to indicate that the interior is too. He thought that this shape was required by his physics (chapter 1), and while that is doubtful, the sphere certainly fits naturally into his world; the Earth is at the center, the planets are inside, and the stars are arranged at the surface of the ball—the edge of space. Beyond that, says Aristotle, is absolutely nothing, 'neither void nor place'. Could our space be like this? Is there some edge out there in every direction?

The arguments against an edge predate Aristotle. Archytas, a friend of Plato's, pointed out an obvious problem: couldn't you stick your hand out beyond it? If that's possible, then whether or not you do so, there is a place outside where the hand could be—a possible place for the hand. But space is the collection of all possible places of things, and so there must be space outside, and hence no edge after all. (Of course, Archytas's argument applies equally *against* a space consisting of the ball minus its surface—an 'open ball'—which is finite but has no edge at all.)

We should notice immediately (as my students point out) that one way around Archytas's argument is to imagine a space in which a bounded space expands whenever one attempts to go beyond its edge. You could stick your hand beyond the current edge, but only because new space would be created, itself bounded. Imagine an ever-expanding version of Aristotle's universe: it is always has an edge, it's just not always in the same place. But that's not the case with Aristotle's space, which has a fixed edge; moreover, an expanding edge is effectively no edge at all. So let's put that case aside and ask whether space could have a fixed edge.

One might imagine an *impenetrable* wall of some kind around a bounded space, preventing hands from being stuck outside. Well, what kind of wall would work in such a case? One could argue that any *material* barrier of finite thickness is penetrable, with enough drilling, or pushing hard enough, say. Then the only kind of impenetrable material barrier is one that is infinitely big. In that case space would have to be infinitely big to accommodate the wall, and one could not have a finite space with a material barrier preventing Archytas from poking his hand out.

However, it's at least logically possible that an impenetrable barrier take the form of a 'force field': for there to be an inward directed force that grows infinitely strong at the edges of a bounded space and prevents escape (assuming all outward directed forces, such as those exerted by rocket engines, are finite). Arguably it's even physically possible. According to Einstein's theory of general relativity, all of space could be inside a black hole, with a spherical boundary at its surface—its 'event horizon'. But black holes are 'black' because gravity inside is strong enough to prevent anything—even light—from crossing beyond its edge (unless quantum mechanics plays a role).

There are also barriers that don't involve forces. Take an example from Henri Poincaré, the late nineteenth century mathematician and physicist (more on him, and the example, in the next chapter). Suppose it were a law of nature that everything shrank smoothly to zero at the surface of a space with the shape of a three-dimensional ball. Then if I briskly walked toward the surface at one step per second, as I got closer, I and my steps would get ever smaller, and I would need infinitely many of them, and so infinitely long, to ever reach there. That is, such a walk will not get me to the edge of space, or my hand beyond it; the ball would thus *appear* to be infinite to those living in it.

(Specifically, suppose that any object shrinks by a factor $1 - w^2$, where $0 \leq w \leq 1$ is the fraction of the way to the edge that the object is; see figure 4.2. That is, when I get halfway to the edge, $w = 1/2$ and I will be shrunk to three-quarters of my original size. My height will be down from 6′ to 4′6″, and my stride from one yard to 2′3″.)

What if I tried moving faster? For concreteness, imagine that I could run as finitely fast as I like, but no faster—any finite (or zero) number of strides per second. By running faster as I approach the edge, I can compensate for my shrinking stride: if, say, I run at twice the number of strides per second at the point at which I shrink to half my original size, then I can keep covering the same distance per second. But I still couldn't escape Poincaré's space: however fast I ran, my strides would shrink so fast that, although I could get as close to the edge as I chose (in as little time as I like), I could never reach the edge itself. The problem is that my strides have no length at all at the edge, which means that no finite number of strides per second will ever even get me there. And if we also suppose that nothing can outrun me, it follows even now that nothing could reach the surface of the finite space.

This case is a bit different from what we considered earlier, for now it is the edge of the space itself that cannot be reached, rather than something beyond. The finite space 'beyond' which nothing can go is here the interior of the ball, Poincaré's space. If only finite paces are possible, then Archytas's argument fails for this space.

The only way to escape from Poincaré's space would be to somehow run infinitely fast. Suppose that I can accelerate so fast that at a finite time in the future I take an infinite number of paces per second. For instance, I could start running at 11 A.M. and continually increase my pace so fast that for any number you like, however large, you could find some moment

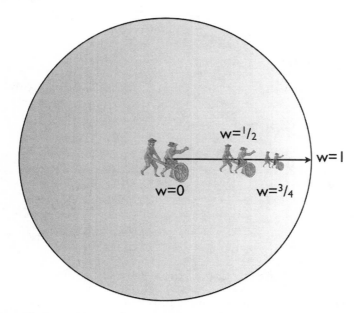

Figure 4.2 In Poincaré's space everything shrinks by a factor of $1 - w^2$ as it moves toward the edge, where w is the fraction of the way to the edge. The two surveyors with their cyclometer or 'waywiser' contract as they approach the edge: halfway ($w = 1/2$) they are three-quarters of their original size (at the center), while three-quarters of the way ($w = 3/4$) they are only 7/16 of their original size. At the edge $w = 1$ so anything that reached the edge would shrink to nothing ($1 - 1^2 = 0$). ('Waywiser' image courtesy Jean and Martin Norgate: http://www.geog.port.ac.uk/webmap/hantsmap/.)

before noon at which I was running at that many strides per second. In the sense that there is no limit to my pace rate, if can run in this way I will run infinitely fast at noon. Such a run is illustrated in figure 4.3 (see the end of the chapter for mathematical details).

The result is that at all times before noon I am located somewhere inside the edge, but at noon (and after) I am not. In a sense, I am on the edge of the sphere, running infinitely fast but stuck, because I have shrunk to a point and my strides cover no distance at all. And so, if such a run is possible, then one could escape the interior of the space. But does this mean that we should conclude, as Archytas would, that the edge must also be part of space?

What's really interesting is that similar reasoning applies to infinite spaces as well (now assuming no shrinking), as the contemporary philosopher John Earman pointed out. Suppose I start running at 11 A.M. and accelerate in such a way that, while I always travel at a finite speed before noon, I have no top speed (again, pick any finite speed; there's some time before noon at which I reach that speed). Then at noon and after

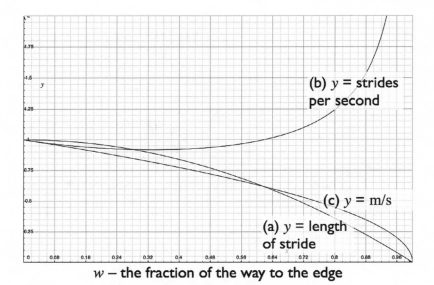

w — the fraction of the way to the edge

Figure 4.3 This graph shows a run plotted against the fraction of the way from the edge, w. Suppose that the runner's stride is 1 m at the center; then (a) shows how his stride shrinks as he moves toward the edge, $w = 1$; (b) shows how fast the runner runs, in strides per second; the rate grows to infinity at the edge, so that at some point on the run the runner runs as fast as you like; (c) shows how taking more steps per second helps compensate for the shrinking stride length; it shows the speed at which the runner traverses Poincaré's space. It becomes zero at the edge, and so the edge cannot be passed. However, this run will reach the edge in a finite time.

I will no longer be in that space! Seemingly paradoxically (though the mathematical point is simple), while traveling at a finite speed at every moment before noon, at noon I would have traveled an infinite distance! (While Einstein's relativity prohibits such motions, there seem to be real examples in Newtonian mechanics.)

The possibility of Earman's run means that Archytas's argument applies equally well to infinite space and to Poincaré's. Suppose you feel that the possibility of a trip taking you out of the interior of the ball implies that it must have space beyond—at the edge. Then, to be consistent, you had better conclude that (prior to relativity) infinite space too must have more space 'beyond infinity' (or technically, 'at infinity'). But that conclusion doesn't seem so obvious. If, instead, you argue that the possibility of Earman's run would not show that there is more than infinite space, then you should accept the possibility of Poincaré's space, and the failure of Archytas's argument even when a finite space can be exited.

That is, the point we need to consider now is whether Archytas's argument works even if there is no effective physical barrier to exiting space, especially since even infinite extent isn't a barrier. What about Aristotle's dictum that existence logically entails being somewhere? One might simply argue that if an object were to reach the boundary (or reach infinity), then it would be nowhere and hence would cease to exist. (Given this possibility, you would be wise to poke a stick through an edge rather than your hand, just in case.)

Or, there is another kind of response. Remember how we understood the way in which the laws of physics determined what was possible—which out of all the possible books of the world were in the 'library' of physical possibilities. Well, if space has a boundary, if we apply Aristotle's dictum, then it must also be logically impossible for anything to exist beyond that boundary. Then there cannot be any books in the library in which something passes through the boundary; imagine starting with all the books in the library of the laws and pulling out those in which anything gets beyond the edge. Such books are as illogical as if they said $1 + 1 \neq 2$, according to the dictum.

What is left? Books where nothing goes near the edge, obviously. But also books in which things go toward the edge but fail to reach it for some reason, despite the absence of any physical barrier: one in which my hand spasms as I try to push it through, one in which I trip and miss the edge, one in which I'm kidnapped as I try, and one in which all those things (and more) happen whenever I try. Such a world would be strange—every attempt to pass the edge would fail, but not because anything in particular is stopping you—but it shows how Aristotle's dictum concerning existence could save his space from Archytas, even without a physical barrier. (In chapter 12 we will see how logic plays a similar role in preventing you from going back in time to change the past.)

After you ask yourself whether space is bounded, an objection like Archytas's probably occurs to you pretty quickly: what would happen if you went past the edge? What we've seen, however, is that there are ways around the problem, both physical and philosophical. (And along the way we learned something about possible accelerations.) Of course that is only to show that space *could* have an edge, not that it does; and absent some good reason to think there is an edge in some particular place, physics makes the assumption that there is none. However, that is not, as you might think intuitively, to say that space must be infinite.

4.2 NEITHER BOUNDED NOR INFINITE

It is a mistake to equate the finitude of a space with the existence of a boundary. For example, consider the spaces in the shapes of the surface of a sphere or bagel. Since they are only *surfaces*, these spaces are two-dimensional: informally, it takes only two coordinates to label each point on the surface (for instance, longitude and latitude for the sphere) and there are only two directions that are at right angles on the surface (for instance, east-west and north-south). In contrast, the space we see and live in is three-dimensional, labeled by x, y, and z coordinates, in the perpendicular left-right, back-forward and up-down directions. (Is three all? We'll discuss that question shortly.)

These spaces also manifestly have finite areas *even though they do not have edges*, because their dimensions are 'closed up' in the rough sense that they join back up on themselves: for instance, if a being travels straight ahead on the sphere, then it will return to where it started, and no curve in either space leads to an edge.

The dimensions of a space need not all be open (like the plane) or all closed up. Consider another simple space: imagine a two-dimensional space that is infinite up and down, but finite—say 1m—across. And suppose that if an inhabitant of this space leaves one side, she instantaneously reappears on the other and so can walk in a straight line back to where she started. Then we can think of the space as having the long sides 'glued' together, so that the ends of her path are joined together. But a strip with its edges glued together is a cylinder, and so the space with the strange rule about passing over edges is really cylindrical space (figure 4.4).

And if you think about it, you can make a doughnut—a 'torus'—by taking a cylinder (of finite height) and 'gluing' top and bottom edges as well. Hence we can think of the doughnut space as a rectangle such that a path across the left side reappears on the right, and a path across the top comes out at the bottom. The area of the rectangle is finite, but because the sides are joined, there are no edges to space. (What about a sphere? In that case all the edges have to be joined at a single point; imagine

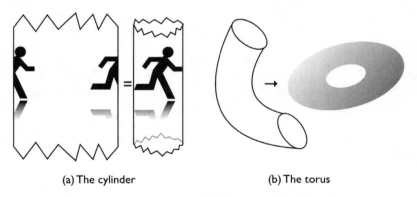

(a) The cylinder   (b) The torus

Figure 4.4 (a) A cylindrical space is a sheet with two sides 'glued together'; put another way, anyone crossing one side of the sheet immediately reappears on the other. (b) If the ends of a finite cylinder are themselves glued together (so that anything crossing one end immediately reappears at the other) then the result is a space the shape of the surface of a doughnut—a 'torus'.

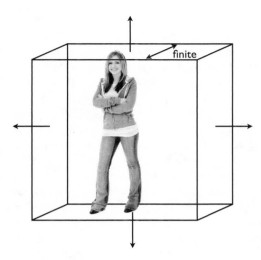

Figure 4.5 This space is three-dimensional. It is infinite in two dimensions—left-right and up-down—but closed up in the third, back-front dimension. The front and rear faces are 'glued together': anything that crosses one of those faces, immediately reappears through the other.

stretching the square into a disc, then wrapping it around a sphere, and joining the edges together at the pole.)

The same is possible in any number of dimensions. It's a little hard to picture, but some or all of the three dimensions we see could also all be closed up, so that one could travel in a straight line back to one's starting point. For instance, imagine a three-dimensional 'slab' of space, infinite in two directions, left-right and up-down, but finite in the back-front direction. And suppose the back and front are glued together. Unfortunately, we can't properly visualize joining the faces in three-dimensions, so it helps to think in terms of traveling around the space. You can travel as far as you like to the left and right or up and down, but if you walk forward or backward, as you pass through one face you reappear through the other, and arrive back where you started (figure 4.5).

Or again, imagine, living in a cube with all opposite faces glued; go across the face in front of you (or above you or to your left) and you immediately reappear across the face behind (or below or to the right)— a three-dimensional torus. (In a three-dimensional sphere, the surfaces are all joined at a point!)

If all the dimensions of space were closed up like that, then our space would also be finite (with the volume of the cube) and yet lack an edge too. And if space is like this, then Archytas's argument does not apply: there simply is no boundary to stick one's hand through. (Bear in mind that Aristotle's ball is not closed up: in that space one doesn't return to one's starting place but reaches the 'end of the road'!)

What is really interesting about this idea is that our space—the one that contains the universe we inhabit—not just some hypothetical space, may really be closed on itself in this way. Indeed, physicists take the question of the topology of space very seriously, and ongoing experiments are attempting to answer it. Currently, there is a proposal that space is a closed up dodecahedron, known as Poincaré dodecahedral space (to be distinguished from Poincaré's ball). A dodecahedron, you may recall, is a 12-sided solid with pentagons for its faces. If space were simply a dodecahedron, then it would have an edge like Aristotle's ball (in fact the media misleadingly described the proposed space as being like a soccer ball). But in a *Poincaré* dodecahedron the faces are glued together in pairs, so that if one exited through one face, one would instantaneously enter through another; imagine a hallway in which every time someone exited left through door 1 (or 3 or . . .) they entered right through door 2 (or 4 or . . .)—the perfect setting for a science fiction bedroom farce.

While the evidence for the closed dodecahedron hypothesis is far from generally accepted (see Further Readings), if it is correct, it would mean that space is finite, even though it had no edge. And that would mean that Archytas was right at least to say that the universe has no edge, but that Aristotle was right to say it was finite, though neither was right for the right reasons.

4.3 WHAT PHYSICS CAN TEACH PHILOSOPHY

In this chapter we have seen how physics—or more precisely, the mathematics at the foundations of physics—can help address philosophical questions. We started with a natural interest in the limits of space, but in order to clarify our interest so that we had some precise question to answer, we turned to physics. First we looked at topology to see exactly how to describe the space: we need to distinguish two- and three-dimensional spheres from open and closed balls; ultimately we saw that finite spaces need not have ends. Then we considered how physics might refute Archytas's argument; what if there were an infinite force or if everything shrank? We saw in the end that something might even 'escape' an infinite space (but to oblivion?). Finally, we have seen how experimental physics can settle the issue.

'Thought experiments' like the ones we considered serve a useful role in physics by directing our attention to strange situations in which certain physical principles still hold. What if forces increased without limit, or if something accelerated without limit? The rules are at least close to those we actually observe, so the examples help us understand what physics allows and how things would change if the laws were slightly different. Such probing of the logical terrain around the borders of a theory lets us understand better what the laws say, and how we might need to change our theories if they come into conflict with experience.

On the other side, our pursuit of Archytas's challenge was not a matter of physics alone; philosophical considerations concerning the nature of existence were also relevant. What if existence requires existing somewhere? Does that mean that things could pass through the edge of space but that they would cease to exist if they did? Or, given our philosophical understanding of laws, should we understand that an edge—but not the laws—would make it impossible for anything to escape? Thus the arguments of this chapter give a nice example of how physics and philosophy can work together on a philosophical issue to the benefit of both.

What we have also done in the last couple of chapters is learn some of the important mathematical ideas on which physics is constructed: something about infinity, about limits, and now about topology. These concepts will continue to be used as we press on.

Further Readings

Some nice explorations of spaces of various topologies (and geometries) can be found at: http://www.geometrygames.org.

Responses to Archytas's argument are discussed in chapter 8 of Richard Sorabji's *Matter, Space and Motion: Theories in Antiquity and Their Sequel*, (Cornell University Press, 1992).

Travels in Four Dimensions by Robin LePoidevin (Oxford University Press, 2003) discusses various aspects of topology and is the source of the idea that Poincaré's space defeats Archytas's argument. (For those who want to know, the runner graphed in figure 4.3 takes $1/(\sqrt{1-w} \cdot (1 + w))$ strides per second. The sense in which the runner reaches the edge is that if we take the product of this function and Poincaré's contraction factor to make sense in the $w \to 1$ limit, then that velocity function will get them to the edge in a number of seconds equal to twice the radius in meters.)

George Ellis's 'The Shape of the Universe' (*Nature* 425, 2003: 566–567) provides a clear, fairly accessible account of the argument for a Poincaré dodecahedron universe. Jean-Pierre Luminet, one of the group proposing the topology describes the ideas in more detail, at a non-specialist level, in his *The Wraparound Universe* (A. K. Peters, 2008). For a review of that book that is critical of the proposal (but otherwise positive) see Andrew Jatte's 'Not Wrapped Up Yet' in *Physics World* (2009, 36).

5

Beyond the Third Dimension?

We've just seen how physics (and the mathematics underwriting physics) can help deal with the question of whether space comes to an end. Now we turn to another topic concerning the 'shape' of space in the topological sense—its number of dimensions. This issue requires even more abstraction than that of boundaries, which we are at least used to. Perhaps when one realizes that geometry in the plane—that is, in two dimensions—is a self-contained subject, one might wonder whether there is some reason that there are three instead. And if one thinks even harder one might wonder whether there is some sense to be made of four (or five or six . . .) dimensions. Indeed, these questions are at least as old as Euclid's geometry.

Aristotle (adopting an argument of the followers of the mathematician Pythagoras) argued that space had to be three-dimensional because the number three represents 'all' in nature. He took as evidence for this claim the fact that when we speak of one thing we refer to it as 'it', to two things as 'both', but to three or more things as 'all' (as in, 'it is such-and-such', 'both are such-and-such', or 'all are such-and-such'). Then how could there be more or fewer than 'all' the dimensions? Well, this line of thought is ingenious, but hardly convincing. It is more or less an accident that ancient Greek and English follow this usage: French, for example, does not, since 'both' is rendered *tous les deux* ('all the two'). The lesson is surely that one should not expect to be able to read facts about the world off language quite so easily.

In fact, we could modify Archytas's argument to try to show that there *must* be more than three dimensions, because limiting the number of dimensions is another way to 'limit' space: for instance, a two-dimensional plane may be only a part of a three-dimensional space. Then the new argument says that if it is possible for anything to move in a fourth dimension (even if it never does), then there must be a possible place for it in the fourth dimension and hence space must have four dimensions. Indeed, one could repeat the argument indefinitely, to show that there are infinitely many spatial dimensions. So unless we can explain how it would be impossible to move in more than three dimensions, then it seems we should conclude that there are an unlimited number of them.

As with bounded space, one could imagine there being a finite number of dimensions, say two, and an infinite force preventing motion in any other dimensions: a force that kept bodies moving only in the plane. However, there is no real reason to think that there is such a force in our world keeping us in three (or however many) dimensions. Even so, Archytas's argument does not succeed, because the laws of mechanics make it impossible to exert any forces in the direction of an extra dimension. We can see how if we assume that space is three-dimensional and consider Newtonian gravity and electric and magnetic forces.

Gravity acts only along a straight line between two bodies, so the force on either body can move it only within our three-dimensional space. And adding more objects won't help: every member of a host of bodies exerts a gravitational force along some line in three-dimensional space, which means that their total effect is again a pull *in* three-dimensional space, not *out* of it. Of course if there were some bodies already off in the fourth dimension, then they could pull bodies out of three dimensions toward them, but that shows only that there could be four dimensions, not that motion in a fourth is possible if there are only three dimensions.

Now, not all electromagnetic forces act along the line between a pair of objects. For example, you may have replicated Hans Oersted's famous (1819) experiment in high school: a small magnet in the form of a compass is placed near a vertical wire, and when a current is passed through the wire, the compass stops pointing north and instead points along the tangent of a circle around the wire, at right angles to a line between the wire and the magnet. (See figure 5.1.) The compass behaves in this way because the current in the wire is nothing but the motion of electrons (in the opposite direction), and the motion of charged particles up the wire causes a magnetic force on the compass, not along a line from compass to electrons, but *at right angles both to the line from compass to electrons and to the electrons' direction of motion.*

This kind of force seems more promising if we want to move something into the fourth dimension: the electromagnetic force is directed out of the plane containing the the wire and the line to the magnet, and into the third dimension. Is there a similar known force that could be used to push something into the fourth dimension? Sadly not. It requires a force at right angles to *two* given directions to push something from the plane into the *third* dimension, so it would require a force at right angles to *three* directions to push something into the fourth, but we know of no such force—certainly not the electromagnetic force. Hence it is impossible for us, given the (known) forces in our universe, to move out of three dimensions, and so the Archytas-inspired argument for additional dimensions fails.

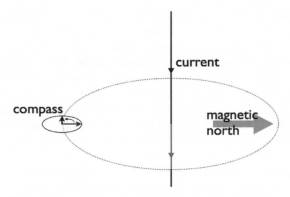

Figure 5.1 In Oersted's experiment, when the current is turned on, the force on the needle is at right angles to the direction of the current (and a line from the wire): orthogonal to the page. When a current flows in the direction shown, the compass needle turns from magnetic north as shown.

5.1 MULTIDIMENSIONAL LIFE

Of course, it doesn't follow that there aren't more (or perhaps fewer?) dimensions than we think, and in section 5.2 we'll consider how and whether we ought to believe that there are—and how we might be mistaken about something so significant. But first it's worth thinking a bit more about what is at stake: we've defined dimensions rather abstractly in terms of orthogonal directions, but what concrete differences would a fourth dimension make? What would it be like to discover a fourth dimension? Imagine, then, that in addition to the familiar three dimensions there is a fourth, perpendicular to it, extending indefinitely, which only a few special people are aware of and can move in. In addition, let us suppose that all everyday objects are three-dimensional and that in the normal course of events we see only things in the familiar dimensions; we don't see light reflected off objects in the fourth dimension. (In the next section we'll consider extra dimensions in which these assumptions don't hold, in which case the following phenomena are not possible. The point here is to use some striking examples to elucidate the very idea of a fourth dimension.)

A rather striking effect is that a fourth dimension could allow solid three-dimensional objects to appear and disappear, and appear to pass through each other. To see this point, imagine perfectly flat creatures living in a plane of three-dimensional space. Since they are confined to a plane, as far as they are concerned space is two-dimensional. Suppose however that one of their number—let's call him Mr. Toody—developed the power of motion through the third dimension (Edwin Abbott's charming novella *Flatland* from 1884 exploits this artifice extensively); he could perform astonishing magic tricks.

over

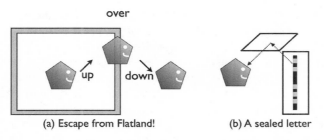

(a) Escape from Flatland!   (b) A sealed letter

Figure 5.2 (a) A Flatlander is trapped in two dimensions by a square, but it could escape if it had access to the third dimension, simply by moving up and over the wall. (b) Flatlanders could use a series of dots and dashes to write letters and seal them in rectangles; Mr. Toody here cannot see in, unless, that is, he gains access to the third dimension, and, say, places a mirror out of Flatland so that he can look right in from the top.

First, he could escape locked rooms. For these creatures, locking him up means placing a closed line around him; if he moves in any direction in the plane, he finds a wall. But he can of course escape such a prison by using his power to move 'up', out of the plane, then over and back 'down' into it, since he has a third orthogonal direction of motion available to him. (See figure 5.2.) Similarly, he could make two-dimensional rabbits disappear and reappear by moving them in and out of the third dimension, and even read a sealed letter by looking into its envelope with a mirror in the third dimension. His feats would appear miraculous to his country-folk, who are accustomed to motion in only two dimensions.

Similarly, if a human had access to a fourth dimension, then she could escape from sealed rooms by jumping out at right angles to our three regular dimensions, pull rabbits out of empty hats from a nearby region in the fourth dimension, and read sealed letters using a mirror in the fourth dimension. In the same way, ghostly apparitions would be possible if spirits could move though the fourth dimension, in and out of our dimensions—and if, of course, there were spirits. (The idea that spirits might be able to move in the fourth dimension is quite old. In his 1671 *Enchiridion Metaphysicum*, Henry More, Newton's teacher, was concerned with what would happen if a soul—which he took to fill a certain volume of space—were to occupy a body too small for it. His answer was that any excess spirit would move off into the fourth dimension, which he called the 'essential spissitude' or thickness.) Again, such things would appear as miracles to us if we tried to understand them in terms of our familiar three dimensions.

Of course we have all seen and heard of such things, though we typically put them down to illusion—whether sleight of hand or camera—delusion, or hallucination. However, if we did find convincingly real examples, then we'd have some reason to believe in a fourth dimension.

I think most right-minded people would be highly skeptical of such 'evidence', but there have been cases in which it has been taken seriously by reputable scientists as evidence for the fourth dimension (which shows that even with the most careful checks and balances all humans can be fooled). For instance, the German astronomer Friedrich Zöllner in the 1870s was tricked by an American magician named Henry Slade into believing that he could tie and untie knots in closed loops of rope, another trick that is possible only in a fourth dimension (see figure 5.3). For a while, his experiences were taken seriously enough to be published in scientific journals of the day.

It is extremely difficult, and perhaps impossible, to visualize directly what is going on in any of the 'tricks' we have described. Of course the problem is that we not only experience everything in three dimensions, we can only picture them in three dimensions. In fact, you may think that we can picture them only in two dimensions, since those are the dimensions of our visual field. However, it is quite clear that, although our visual experiences are of the two-dimensional surfaces of objects, they also have a sense of depth: we experience objects in three-dimensional space. (For one thing, we have two eyes, receiving slightly different images, from which varying distances can be computed. There are other visual cues as well, such as lighting, that we use to judge distance.) Not only that, we are able to move and rotate objects in three-dimensional

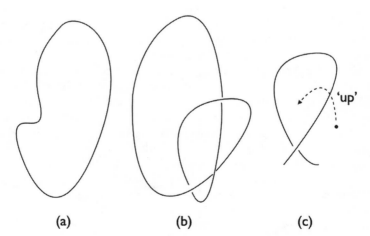

(a)   (b)   (c)

Figure 5.3 It is impossible to take a loop of string (a) in three dimensions and tie a knot in it (b). However, it is possible in four dimensions. (c) A plane cross-section through the knot; the string that is to run though the middle of the loop (the dot in the picture) runs orthogonal to the page in three dimensions, so it cannot be lifted in that direction. But if space has four dimensions, then there is another dimension orthogonal to the plane shown, and the string can be lifted 'up' and through in that direction.

space in our 'mind's eye', to see what they would look like from different angles and so imagine what volume they occupy.

Going beyond three is difficult; those faculties don't allow us literally to picture four-dimensional space. That of course doesn't mean there couldn't be such a thing. After all, we can imagine Mr. Toody being in an analogous situation. Suppose that he has a one-dimensional field of vision; we see the two-dimensional surfaces of three-dimensional bodies, while he sees the one-dimensional edges of two-dimensional figures. But suppose, like us, he has a sense of depth and the ability to move and rotate objects in two dimensions in his mind's eye. That is, he can visualize two dimensions just as well as we can three. But these faculties will fail him when he tries to think in three dimensions—to think, that is, like us. He has to be content with partial images and understanding the mathematics of the third dimension; we have to be satisfied with the same approach to the fourth.

One way to help understand an additional dimension is to use time as a stand-in; we are well equipped to imagine processes over time. To imagine moving in the fourth dimension, you have to imagine having the power to travel in time as you pleased (the topic of chapter 12). For instance, you could pull a rabbit out of a hat by placing it a little further on in the 'fourth dimension', at the same place the following day, say. Just quickly travel in time, retrieve it, and pull it out of the hat.

Or (to use an example from Isaac Asimov's short story *Gimmicks Three*), you could escape from a locked room by traveling backward in time—in the 'fourth dimension'—to a point before the room was built, walking outside the future location of the room, and traveling forward through time back to your starting point. Just as you find no walls in the past, you find none in the fourth spatial dimension. (There's a bit of a disanalogy here: a three-dimensional room exists only at one point of the fourth spatial dimension, just as two-dimensional walls exist only in a single plane of three-dimensional space. In the dimension of time, the walls first grow and then persist; the wall is four-dimensional if the fourth dimension is time.)

5.2 MORE THAN THREE DIMENSIONS?

Is all this talk just fairy tales? After all, we do experience only three dimensions. Are there any reasons for believing in more? As in all cases, we can take the immediate evidence of our senses to be reliable only if we have no overriding reasons of indirect experience to doubt them: for instance, tables appear solid, but there is plenty of evidence to think that they are really composed of atoms in space. Is there any similar evidence for the existence of other than three dimensions? Are there any reasons for thinking that there are extra dimensions of which we are not aware?

Astonishingly the answer is 'yes', but unfortunately not in a way that is likely to permit any of the magic tricks we considered above (again, those tricks are meant to help us picture extra dimensions). This positive answer comes from string theory. It's important to bear in mind that this theory, though the focus of much interest, and taken very seriously by many physicists, is still highly speculative, and so are any of its consequences. One of those consequences is that space must have far more than three dimensions—at least nine (plus the dimension of time), depending on the version of the theory. One might think that string theory is then rather obviously refuted by our immediate experience, but there is a way out: if string theory is right, then most of the dimensions must be very small, and only three are large enough for us to be aware of them!

To see how such a thing is possible, recall our earlier discussion of cylindrical dimensions (see figure 4.5). Imagine a three-dimensional space, infinite in two directions but cylindrical in the third. A three-dimensional being—you, say—in such a space can move off as far as it likes in two directions, but in a third direction after a certain distance will simply return to where it started. In other words, space is like an (infinite) slab, with the two faces 'glued together', or 'compactified' to use the standard jargon. It requires extra dimensions to do the joining literally, and so it's hard to picture; so instead we think in terms of the rule that things leaving one face immediately reappear through the other. Hence, a ball thrown in the finite, cylindrical direction will hit us in the back, and supposing light to behave in the same way, if we look in that direction we see the backs of our heads in the distance.

Now imagine that the distance around the cylindrical dimension— the distance traveled before one reaches one's staring point—starts to shrink. You see yourself come closer; as the dimension shrinks, the distance between your front and back decreases, until your back starts to press into your front. Now the circumference of the dimension equals your thickness, and if the dimension shrinks further you are simply crushed to fit into it. Rather sad for you, but imagine three-dimensional creatures adapted to a three-dimensional cylindrical space with a tiny circumference, $1/1,000,000,000$ meter, say. Such beings could be any size in the 'large' dimensions—human sized, say—but no bigger than $1/1,000,000,000$ m in the small 'compactified' dimension (see figure 5.4).

And now imagine a four-dimensional being adapted to fit in a four-dimensional cylindrical space, in which the compactified fourth dimension is even far tinier. That's you, according to string theory (if we focus on just one of the six or more small dimensions).

Clearly the size of the small dimension gives some intuitive appeal to the idea that it's too small to notice. However, we can add a little more to the story to see why that should be the case according to string theory. The short answer is that compactified dimensions are too small

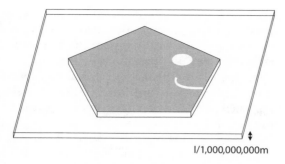

1/1,000,000,000m

Figure 5.4 A three-dimensional cylinder again, but now with a tiny cylindrical dimension, only 1/1,000,000,000 m across. Although this creature has three dimensions, one is too small for him to notice.

for anything much to happen in them, and so almost all of the time all the action is essentially three-dimensional.

Just consider what's happening around you, and consider describing what happens in a very thin horizontal slice, extending 100 m all around but only 1 mm thick. Of course you are going to find that there are all sorts of variations in the horizontal directions that need to be described: table here, air there, a wall in that direction, a roof in that. But typically there aren't going to be many differences in the vertical direction at any point that are worth mentioning; the table is very constant across 1mm and so is the air.

And the thinner the slice, the more the point holds; if the slice were 1/1,000,000,000 m (or 1/1,000,000,000,000,000,000 m) then there would be essentially nothing to say about what was happening in the vertical dimension. So, once you've described the two-dimensional, horizontal layout of your three-dimensional slice, you've said it all; you might as well be describing a two-dimensional region.

But a small, closed-up dimension makes space just like this; there is a lot of variation across the large dimensions but effectively none across the compactified one. Hence, while things actually happen in all dimensions, the pattern of events occurs as if it were occurring in a space without the compactified ones; the descriptions are the same. But what we are aware of is not space directly, but the things that happen in it; they happen in many dimensions, but they happen as if they were occurring in three, and so they appear to happen in three-dimensional space. It's worth pausing to think about: if string theory is right, then contrary to appearances, everything—you included—has many dimensions and everything happens in many dimensions. It just doesn't look that way!

At least it doesn't look that way unless you look very, very hard, in situations in which you make something happen in the fourth dimension. So the situation is a familiar one; things appear one way and then you find compelling indirect reasons to suppose that they are otherwise, and you can then explore them more directly. For instance, everyday objects such as tables, chairs, and floors appear solid, but in fact, according to atomic physics, are mostly vacuum with small atoms tightly bound together. If we look hard enough, though, we can see the atoms. Any reasons for giving credence to string theory are reasons for believing in more dimensions than four; the problem is that direct experiments on the strings are far beyond our current abilities.

One last point, which helps highlight the fluid state of the current physics of space. According to the 'holographic principle', a conjecture to which a number of physicists have devoted considerable attention, the physics in a region of space may be 'projectable' onto the surface of that region; all the information about the state of the region and its evolution may be captured by the state and evolution of its surface. In that case, even when a space appears to be n-dimensional, it may in fact be $(n - 1)$-dimensional (the dimensionality of the surface of an n-dimensional space)—we may mistake the physics of some space for the physics of the space it surrounds! Clearly the question of dimensions is a live one.

The point of this chapter was to explain the meaning and significance of the dimensions of space. We've seen both how to think about dimensions and why contemporary physics gives us reasons to believe in them. We discussed the possibility of reviving Archytas's argument, but aside from that our main concern has been to understand the physics of dimensions rather than the philosophical issues that they raise. With this understanding behind us, we can now turn to some philosophy.

Further Readings

Michio Kaku's *Hyperspace: A Scientific Odyssey through Parallel Universes, Time Warps, and the 10th Dimension* (Anchor, 1995) is a wonderful discussion of modern physics, with particular emphasis on spaces of many dimensions (and string theory), including the popular and scientific history of the idea. He tells the tale of J. C. Friedrich Zöllner, whose "On Space of Four Dimensions" appeared in *The Quarterly Journal of Science* 8 (April 1878).

6

Why Three Dimensions?

Our immediate experiences reveal a world of three dimensions: the three orthogonal directions, up-down, left-right and back-forward. Perhaps these are all that exist, or perhaps string theory is correct and these are just the 'large' dimensions. Either way, is three an arbitrary number, or is there some reason for it? We rejected Aristotle's argument in the previous chapter, but now we will consider another, more influential, line of thought, which connects Newton's law of gravity to the dimensions of space.

6.1 THE FORCE OF GRAVITY AND THE DIMENSIONS OF SPACE

As an analogy to gravity reaching out from a massive body such as the sun, consider the fragments of an exploding bomb. The shrapnel will form an expanding sphere around the place where the explosion occurred (figure 6.1).

The area of a sphere increases as the square of its radius: twice the radius means four times the surface. However, the total quantity of shrapnel remains the same as it flies out, so it must be spread out in an increasingly diffuse sphere. When the radius of the sphere doubles, so that the area covered by the shrapnel quadruples, the amount of shrapnel per unit area—its density—must be one quarter of what it was. In other words, the density of shrapnel decreases as the inverse square of the distance. (The same story applies to light expanding from a bulb, and hence the futility of taking flash photographs beyond a few meters— the amount of reflected light returning drops off as the square of the distance.)

In rough terms, one can think of gravity as a fixed 'quantity of attraction', propagating analogously to shrapnel, diffusing over the surfaces of increasing spheres, with the force at any point depending on the 'density of attraction' at that point. Since density varies as $1/r^2$, one expects on this model a $1/r^2$ law for gravity—as we indeed find.

This line of thought depends, however, on the dimensionality of space. In n-dimensional space, concentric 'spheres' are $(n-1)$-dimensional and

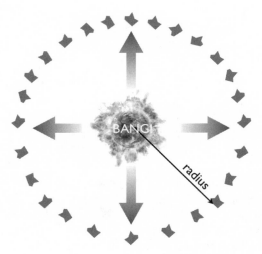

Figure 6.1 Pieces of shrapnel from a bomb form a sphere which expands away from the explosion; here we see the cross-section. The surface area of the sphere (like that of any sphere) is $4\pi \times$ radius^2.

increase in size as the $(n - 1)^{th}$ power of their radii. In a two-dimensional plane, fragments flying in every direction from an explosion form concentric circles (one-dimensional 'spheres'), which of course increase in circumference in proportion to their radii (recall, circumference = $2 \times \pi \times$ radius). In three-dimensional space, spheres are two-dimensional with areas proportional to their radii squared. Similarly—though it is not readily visualized—concentric 'spheres' in four dimensions are three-dimensional, and their volumes increase as their radii cubed: twice the radius means $2^3 = 2 \times 2 \times 2 = 8$ times the volume. And so on.

Imagine now an explosion in a space with an arbitrary number of dimensions, n, with the shrapnel spreading out in an expanding sphere of $n - 1$ dimensions. Since the total quantity of shrapnel remains the same as it expands, but is spread out over a sphere whose size increases as $r^{(n-1)}$, its density decreases like $1/r^{(n-1)}$. In two dimensions, if there is 1 kg of shrapnel per meter around a circle, then when the shrapnel circle has grown to twice the radius, there will be 1/2 kg per meter. We've already seen the case for three-dimensional space, and for four-dimensional space the density of shrapnel would decrease as $1/r^3$—twice the distance means $1/(2 \times 2 \times 2) = 1/8$ the density. And so on.

So according to the analogy with gravity, we should expect the force of gravity in n-dimensional space to diminish as $1/r^{(n-1)}$: as the quantity of attraction 'spreads out', it gets weaker, in just the way shrapnel gets more diffuse. We indeed find this result for our three-dimensional space: we observe gravity to follow the $1/r^2$ law.

The idea that the form of gravity is connected to the number of dimensions originates (as far as I am aware) with the giant of philosophy, Immanuel Kant (1724–1804), who made the claim in a 1747 essay entitled *Thoughts on the True Estimation of Living Forces*. He gave no detailed argument for the connection, and one might be doubtful about it. Indeed, you might well be dubious about the argument from analogy: is gravity anything like a fixed quantity of shrapnel spreading out though space? Couldn't the law of gravity be, say, $1/r$ or r^2 in three dimensions, or $1/r^2$ in two or four or whatever dimensions? Such forces might not be as intuitive, but is it impossible for the law to take those forms? After all, nuclear forces don't obey the $1/r^2$ law, and we have long learned that nature is no great respecter of our intuitions.

In Newtonian mechanics, this objection seems cogent; his laws of motion are compatible with other laws of gravity. However, according to the laws of general relativity, if space has three or more dimensions, then it does follow that far enough away from a body like the sun, the gravitational force must obey a $1/r^{(n-1)}$ law, as an approximation that gets worse as one gets nearer to the body. Note, however, that two dimensions is a special case; there is no gravitational force toward massive bodies. And so all this talk about shrapnel should be understood as a nontechnical analogy to the derivation of this result in general relativity (one that does not work in two dimensions).

It's well worth noting that we have found a new way to observe extra dimensions: by measuring the form of the law of gravity! One has to be careful, of course. From the moon's orbit, Newton calculated a $1/r^{2.016}$ law, so should we infer 3.016 dimensions? (That's not utterly absurd; physicists have some understanding of what fractional dimensions would be like.) No, he put the difference down to the effect of the sun on the moon's orbit, arguing for $1/r^2$.

More relevant to our discussion is the possibility of using gravitational measurements to observe the existence of small compactified dimensions. For the small dimensions count only at distances smaller than their circumference; a sphere of shrapnel can expand only so far in a compactified dimension, after which the area increases only because of expansion in the large dimension. Thus Newton's law holds for the planets because the solar system is much larger than our compactified dimensions (if we really have them).

So to search for the extra dimensions, one tries to observe deviations from Newton's law at short distances. How short? We said above that string theorists estimate them to be extremely small by any standards, but what is really interesting and exciting is that until recently no one had performed such experiments to the limits allowed by current technology. People realized that we might be able today to observe extra, compactified dimensions if they were some reasonable fraction of a millimeter! So far, however, such experiments have not revealed deviations from the $1/r^2$ law over 1/10,000 m; if there are extra dimensions, they are smaller than 1/10,000 meter. Experiments continue.

6.2 DOES INTELLIGENT LIFE TAKE THREE DIMENSIONS?

Now we are in a position to consider how a number of physicists and philosophers have seized on the purported connection between the form of gravity and the dimensions of space as part of an explanation of why space has three dimensions. These arguments can be traced to a 1955 essay by the mathematician and philosopher G. J. Whitrow (arguments repeated and elaborated by Stephen Hawking in his famous *Brief History of Time*). Whitrow's argument has three main steps; each is well worth considering, though I will explain why they are problematic.

First, on the basis of the gravity-dimensionality connection, Whitrow argues that intelligent life cannot exist if space has more than three dimensions. Second, he argues, on other grounds, that one and two dimensions are incompatible with intelligent life. The final step is to argue that since intelligent life requires that space have neither more nor fewer than three dimensions, we can reasonably explain the dimensionality of space on that basis. This final step is especially perplexing—space has three dimensions *because* there is intelligent life?!—so we will discuss it in more detail later; but let's look at each step in turn.

In his essay Kant pointed out that it is possible for space to have other than three dimensions, provided that the law of gravity takes the right form. So you can explain why space has three dimensions on the basis of the law of gravity only if you first explain why the law of gravity itself takes the $1/r^2$ form. Kant himself believed that gravity could obey any law, so that space could have any number of dimensions. Other thinkers, however, have tried to show that the form is highly restricted. Specifically, they argue that intelligent life is not possible if gravity takes a $1/r^3$ or $1/r^4$ form or any higher inverse power, and hence that intelligent life necessitates that space have at most three dimensions.

What does intelligent life have to do with gravity? As Paul Ehrenfest showed in 1917, if the planets and sun had mutual attractions that decreased like $1/r^3$ or as the inverse of some higher power, then small collisions or wobbles would cause planets to fall out of their orbits and either collide with the sun or fly away. So only the $1/r^2$ and $1/r$ forms of gravity allow for the prolonged existence of solar systems. But, Whitrow and Hawking argue, it is likely that stable solar systems are a prerequisite for the evolution of intelligent life; it takes a long time in a temperate corner of the universe for intelligence to evolve. Hence intelligent life can arise only if gravity has the $1/r$ or $1/r^2$ form, and so, if gravity and dimensionality connect as Kant claimed, intelligent life can exist only if space has two or three dimensions.

The second step of Whitrow's argument is to argue that two-dimensional space would not allow intelligent life at all. His (and Hawking's) argument is that two dimensions do not permit the kind of digestive system required for life (similar reasons would rule out one dimension). Our digestive system is, crudely speaking, a long, irregular

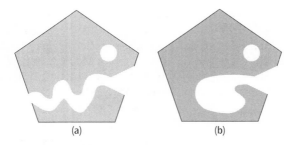

Figure 6.2 (a) If a two-dimensional being like Mr. Toody has a digestive system like ours, then he is in danger of falling in half! (b) But if he eats and excretes through his mouth, there is no problem!

tube running from mouth to anus. Since we are three-dimensional, our bodies entirely surround this tube and so have integrity. In two dimensions, such a tube would amount to a channel cutting right through a plane figure, dividing it in two, with nothing to hold the two halves together; if food could pass through Mr. Toody then he'd fall apart! (See figure 6.2.)

This argument is flawed in a number of fairly obvious ways: first, there are many life forms in our world that don't have a digestive system like ours—single-celled organisms, for instance. Whitrow recognizes this objection and proposes that intelligent life at least requires a gut. But even if we grant this point (and why should we?) the argument fails. For one thing, one could imagine a digestive system in which a two-dimensional creature was always careful to shut its mouth (and upper portion its digestive tract) whenever it excreted; no problem with falling apart then. Or a two-dimensional being might, like the admittedly unintelligent jellyfish, swallow her meal, digest it in her stomach, then simply regurgitate the waste; not very nice, but better than coming apart at the seams. Without further argument, there is no reason to think that intelligent creatures couldn't have such digestive systems, and indeed that they might evolve to have such systems in a world of two dimensions.

A more promising line of attack for Whitrow's conclusion might be to argue that one cannot construct a complex brain in two dimensions. Computer circuits are three-dimensional (if fairly thin) because they involve wires crossing over each other. If a brain is sufficiently like a computer, then perhaps one cannot be built in two-dimensional space, and so two dimensions would preclude intelligence.

However, we know that computers can be built in two dimensions, because of work done by mathematicians on 'cellular automata', such as John Conway's 'Game of Life', which we discussed in the Introduction. Interpreting alive creatures as 1 s and dead creatures as 0 s means that computer data—which is nothing but a pattern of 1 s and 0 s—can be stored in the grid and then evolved (according to the laws of the game), which is in effect exactly what a computer does. Mathematicians have shown that such worlds can perform any calculation that a computer of any complexity can (though of course the method of input and output is cruder; there's no keyboard and no monitor). So we have the remarkable result that any computation at all can be carried out in two dimensions (and, in fact, even more remarkably, in one dimension)! And so computer architecture is no reason to think that intelligence is impossible in two dimensions.

One might now want to argue that computers aren't intelligent, but then why should the number of dimensions have any bearing on the possession of whatever 'extra' it is that intelligences have that computers don't? A more reasonable response is to argue that it is far less likely for two-dimensional than three-dimensional brains to evolve: because, perhaps, an intelligent cellular automaton is too large or too slow to have much chance of survival. Maybe such an argument can be made, but I've never heard one.

Thus, if we want to argue that intelligence is incompatible with two (or one) dimensions, then the argument will have to be similar as for four or more dimensions: general relativity does not permit solar systems. Recall, contrary to the shrapnel model, general relativity does not entail a $1/r$ law for two-dimensional gravity. If it did, then it would be possible for two-dimensional planets to orbit a two-dimensional sun in a two-dimensional space. But as it is, according to general relativity, there are no stable solar systems in fewer than three dimensions, just as there are none in more than three.

6.3 IS THE UNIVERSE MADE FOR HUMANS?

Thus unless you assume general relativity, there is no compelling reason to connect dimensionality to the existence of intelligent life. If, however, one does assume general relativity (a very reasonable assumption) and if one assumes that intelligent life can arise only via evolution on a planet in a stable orbit around a sun (a reasonable but less certain proposition), then one can conclude that three-dimensional space is necessary for intelligence. But so what? According to Whitrow and Hawking, one can thereby *explain* why space has three dimensions. But how is this third step of their argument supposed to work?

This question is especially interesting because it invokes a remarkable style of (philosophical) argument—'anthropic' reasoning—that has become rather popular among physicists working on the quantum physics of space. The problem that they face is that typical quantum effects occur on such mind-bendingly small scales that there are really no experiments giving clear guidance in the construction of a theory: few experimental facts ruling out one idea rather than another. One of the few relevant empirical facts available is that we live in a universe that supports human (that is, *anthropic*) life. Anthropic argumentation covers a variety of attempts to milk this information—for example, to argue against theories that make life unlikely. Whitrow and Hawking employ another variety of this reasoning.

They claim, on the assumption that intelligence requires three-dimensional space, that there are three dimensions *because intelligent, human, life exists*. One naturally feels that this is not the kind of explanation one was hoping for at all. I certainly feel that way. When we ask why there are three dimensions, since there is no mathematical necessity for it, what we want to know is for what *physical* reasons are there three dimensions: *given the laws and forces of nature*, are three dimensions necessary?

Spelling out our question in this way makes an important point about 'why-questions': they come with (often unstated) presuppositions concerning what is to be taken for granted, so that one and the same sentence—such as 'why are there three dimensions?'—may be used to ask a number of different questions. For example, the contemporary philosophers Nuel Belnap and Bas van Fraassen suggest that all 'why' questions demand an explanation of why one thing happened in contrast to another: 'why X rather than Y?' or 'why X rather than Z?' However, we usually take the contrast as read, and so just ask 'why X?' and let the listener figure out which of the two questions '...rather than Y?' or '...rather than Z?' is being asked. Typically, we use social and linguistic conventions, and by the context a listener has no trouble in figuring out the correct contrast: for instance, if asked 'why are you wearing a fur coat?' on a hot day, we'd take the contrast to be 'rather than no coat at all', but at an animal rights meeting the contrast would be 'rather than something ethical', and in each case we would expect different explanations.

We can shed some light on Whitrow and Hawking's explanation by giving this kind of analysis: by stating explicitly the relevant contrasts and other presuppositions that distinguish some of the questions one might be asking when one asks why space has three dimensions.

One might conceivably take the question to mean 'why do humans live in three-dimensional space rather than a space of two or four or more dimensions?' to which Whitrow can answer, 'because humans can live only in three dimensions'. This rewording of the question makes clear the assumption that humans exist, and it really asks what it is about three-dimensional space that makes it especially suited to human life, to which Whitrow's response seems reasonable.

However, if one asks, 'why are there three dimensions?' what one probably means to ask is simply 'why three, logically or physically, rather than some other number?' with no reference to the existence of humans. And in response to that question, it is simply irrelevant to point out that humans could not exist in other spaces, for why should humans exist? In other words, Whitrow's whole approach ultimately misses the point: it seems to be the wrong kind of explanation because it doesn't address the question that we want it to.

Hawking answers yet another version of the question. First he reinter-
prets it to mean 'why is space *observed* to be three dimensional, rather
than some other number?' and points out that this question implicitly
refers to someone observing the dimensionality of space, that is, to intelligent life, particularly us. Analogously, we might ask why we observe our
surroundings to be the surface of the Earth rather than the bottom of the
ocean or deep space. Now one naturally takes this question to mean 'given
the existence of the surface of the Earth, among other places, and given
the existence of humans, why do we observe the Earth's surface around
us rather than some other surroundings?'; to which it seems perfectly
reasonable to say, 'because the surface of the Earth, unlike other places,
is convivial to human life'. Following this kind of model, Hawking takes
his question to mean, 'given the existence of three-dimensional regions
of space, among others, why do we observe the region we inhabit to be
three-dimensional rather than some other number?' Analogously, if this
is a reasonable question, and if only three-dimensional space can support
intelligent life, it is reasonable to answer with Hawking, 'because only
three-dimensional space allows intelligent life'.

But again, this is not the question that we thought we were going to
have answered when we asked why space has three dimensions. Indeed,
Hawking's question assumes that we already have an answer to the
question we wanted answered, because it assumes that there are three-
dimensional regions of the universe! What we expected to hear was how
logical or physical principles somehow required (or made likely) that
state of affairs, and so getting an answer to Hawking's question is again a
disappointment.

Hawking realizes the crucial presupposition of his question and claims
that physics—by which he means string theory—allows the existence of
three-dimensional regions of space (among regions of other dimensions).
However, the mere 'possibility' of regions of different dimensionality is
not sufficient for the kind of answer Hawking proposes; we really need
some independent grounds for thinking that they are there.

6.4 THE MEGAVERSE

In a recent book, *The Cosmic Landscape*, Leonard Susskind describes what
he (and a number of other prominent physicists) take to be such grounds.
The first point is the observation that many, many universes, with very different physics, are compatible with string theory. The imaginary 'library'
of string theory has many books, with many possibilities, corresponding
to different arrangements of matter, different kinds of matter, different
kinds and strengths of forces, and different geometries—and different
dimensions.

Metaphorically, what Susskind proposes is that string theory also allows
'megabooks' which describe universes in which different regions of the

universe are described by very different books, with different kinds of matter, forces, geometries, dimensions, and so on. A megabook describes what Susskind calls a 'megaverse', while the smaller books describe 'pocket universes' within it. An important part of Susskind's picture is that each of the pocket universes is expanding like a bubble, so you can't get out of one, but the whole megaverse also expands, so the bubbles don't collide.

Much as we discovered that there was much more to the universe than Aristotle imagined, so now we are to suppose that there is even more to it than we have imagined up to now.

So far the picture sounds just like Hawking's idea that string theory allows regions of different numbers of large dimensions. However, Susskind is able to add a considerable amount of important research on the precise nature of the different universes. Most interestingly, what he is also able to add is the claim that if a universe starts out in one of the ways allowed by string theory, then there is a small, but not vanishing, probability that any point will 'jump' into another possibility. Then a pocket universe of the new type will grow from the point.

This proposed process occurs very slowly, but it occurs everywhere, even in new pockets, so over time most possibilities get tried out somewhere, and so most kinds of world occur somewhere, producing the megaverse. If that's right, and if Hawking and Whitrow are right that intelligent life requires there to be three large space dimensions, we see only three because that's the only kind of pocket in which we could live (and similarly for a number of other physical quantities). And that would be a perfectly reasonable anthropic explanation of the dimensions of space.

So is this the explanation we were seeking? At best it is far too early to say. There are two kinds of difficulty: first, the physics I have described is speculative, even by the standards of string theory; and second, even if it could be filled in, it is not clear why we should accept it. (To be fair, Susskind acknowledges these difficulties and tries to confront them.)

First, why should 'jumps' to other universes occur? The idea is that quantum mechanics, being indeterministic, will allow a system to leap spontaneously from one state to another with some probability. That is what happens in radioactive decay, for instance: an atom spontaneously jumps from one state to one of lower energy, emitting radiation in the process. (For instance, a uranium atom can decay to thorium, emitting an 'alpha particle' of two protons and two neutrons.) But the idea is more general: if there is another state out there, then there is usually some probability of reaching it. So the idea of the megaverse is that since other states—other possible pocket universes—are 'out there', quantum leaps will eventually get to them.

The problem is that the standard quantum understanding does not apply to the formation of pocket universes, because of the way they expand (roughly speaking, too fast to make sense of the connection between the megaverse and its pockets). Susskind thinks that the resolution will require some kind of modification of quantum mechanics, and he has been working to develop it. (The 'holographic principle', mentioned earlier, is a big part of his idea.)

Neither I nor (likely) you are well qualified to judge the likelihood of success—certainly not as well qualified as Susskind! But while the idea is far from a completely worked out theory, let's suppose that he's right. Even so, there is a problem with accepting it as a realistic model of physics.

Suppose that there really are all the different possible string universes that Susskind claims, and a theory of their 'birth' within a megaverse. According to the proposed view, we are living in just one pocket. The problem is that given everything we know about the universe or are ever likely to see, it is also possible that we are not in a megaverse but that the pocket is everything, and always will be. By the design of Susskind's model, the physics around us will be just the same either way; the pockets are just little universes.

According to the megaverse model, pockets will start forming, but that is not required by the string physics of the universe. Pockets form only if we add the assumption that other universes are *dynamically* accessible, the kinds of states out there that can be reached by some physical process. But it does not follow from the mathematical cogency of a state that it is dynamically possible.

The situation is much like that we confronted when discussing Archytas's argument. To try to show that the universe couldn't be a closed ball, Archytas argues that there are locations on the outside that could be reached—that are dynamically accessible by ordinary motions. But one could respond that while points beyond the boundary are mathematically possible (because an infinite space makes mathematical sense) that fact doesn't imply that they are dynamically accessible at all. As we suggested, maybe the topology of space is part of the laws so that no books with bodies passing through the boundary are allowed. Weird things might happen at the edge whenever a body reached it, but it's a possible scenario. And in that case a mathematical possibility does not after all imply a dynamical one.

And in the case of the megaverse, things are, as far as I can tell, even weaker. There is a sense in which the topological law is artificial, put in to stop otherwise natural motions out of Aristotle's universe. In the case of string theory there is no need to impose a restriction on possible processes to prevent the formation of pocket universes, for they have to be added to string theory in the first place. That is, string theory by itself does not allow for the natural growth of pocket universes.

I'm sure Susskind is sensitive to this point, but what are needed are some reasons to think that the extra processes should be added to our theory. He discusses how influences from other pockets might reach ours, so the problem is not that there is no possibility ever of getting experimental evidence. The problem is whether anything now gives us a reason to believe in the megaverse.

For Susskind it is the existence of humans. For instance, there seems to be no other explanation for three dimensions than the anthropic one involving the megaverse. Well, being part of a good explanation is a good reason to believe in something. For example, why do we believe in atomic particles? Because they explain so much. But in that case there are a lot of independent ways in which the particles enter explanations, giving a range of reasons to believe in them. But for the megaverse we have just humans, and the various conditions necessary for human life. It would be nice to have some other physical reasons to accept the theory.

But as I said, the physics in any pocket is effectively indistinguishable from the physics of a nonpocket universe. So the situation is quite unlike that facing Newton, for instance, when he saw that gravity and the force on the moon were one and the same thing. We don't see our universe and another and realize that through string theory they are parts of the same megaverse. We just see ours and postulate others.

In both cases the reasoning is by *induction*, inferring something more general from specific observations. Newton saw how bodies fall and orbit, and he showed how they could be examples of a single force surrounding the Earth, described by a single law. The induction there is grounded in very concrete physical properties of things. In the case of the megaverse, things are much more tenuous: from the physics of our universe we learn that string theory may be true, and then we realize that the megaverse is possible. But that leaves us a long way from having reasons to believe it that are anything nearly as strong as Newton's.

I want to point out a philosophical issue in the background of this whole debate. Susskind starts with string theory and its many possible universes—the universes compatible with its laws. Then he creates a new law which will allow a universe to make a quantum 'leap' to another, leading to its 'birth', and so on, to create the megaverse. But quite a step is involved here, from finding a universe in the library of string theory, to arguing that there must be a megabook linking all those books, that all the books describe reality.

Consider, for instance, the account of laws that I sketched earlier, that they are just comprehensive and concise descriptions of what happens in our universe. Then what does it mean for a book to be in the corresponding library? Just that it describes a universe compatible with our favorite description of what happens in ours. But that doesn't seem to give any reason to think that our universe can make a quantum leap to it; to see the second universe as possible in that sense would be to forget that it is just in the library picked out by our describing things a certain way.

So I suspect that in the back of his mind Susskind has a different picture of what it is for a universe to be possible, so that quantum leaps are always allowed between universes in the library of string theory. Somehow, the library isn't determined just by what happens in our universe, but is given by nature as a whole. Then there is a sense that we are just one among many predetermined possibilities, which are 'out there' to be reached.

I don't mean to settle the issue here but rather to point out that philosophical views about laws and the meaning of possibility color one's views. The bottom line is that although Susskind has certainly advanced the idea of an anthropic explanation both in details and justification, there is a long, long way to go before we can happily accept his megaverse.

6.5 PHILOSOPHY IN PHYSICS

In this chapter we've seen that the relevance of philosophy to physics is not just something in the past, like Zeno's paradoxes. Contemporary physics, in the form of string theory, has given life to questions concerning space's dimensions. On the one hand, it provides new reasons to believe in them, and new models of how they relate to the familiar three dimensions. On the other, physicists have been led to use the fact of our existence as a data point in constructing theories and explanations.

Such an idea sounds like a step back to the Aristotelian picture of things, with humans taking a special place in the universe. But that is not the intent, at least. Intelligence seems to be a rather unique characteristic, both in our universe and according to the laws of physics, so it is natural to consider its significance.

However, compared with most experimental data appealed to by physics, it is very singular. Usually one likes to be able to see how one quantity varies as another is changed: How does the position of the moon vary over time? How does the boiling point of water vary with the amount of salt dissolved in it? How does the form of gravity vary with distance? With such information, one can start to understand what kinds of laws might explain such correlations. But intelligence is not like that, it's just here. In this case, it takes philosophical reflection on how physics relates theory to experiment, and how physics explains, in order to understand how to properly use the fact of intelligence. We have indeed seen some of the pitfalls to be avoided.

Additionally, intelligence is not usually thought of as falling within the subject matter of physics; the question of the connection between physical entities such as brains and computers and intelligence is philosophical. And so again, the arguments of the kind we have been looking at bring philosophy into physics.

In fact, most contemporary philosophers (certainly) and physicists (I'm pretty sure) would accept that a brain or computer is capable of intelligence, simply in virtue of its physical properties. At the very least, they are going to assume, as in this chapter, that the physical properties of a person have an important effect on what happens in his or her mind. Just how, is of course arguable: for instance, I have just argued that the absence of a digestive tract has no bearing on intelligence. However, the idea that physics constrains us, and especially our minds, in philosophically relevant ways will become increasingly important as we progress.

Further Readings

The experiments to measure the form of gravity at distances less than 1 mm are described in *Physics Today Online*: http://www.aip.org/pt/vol-53/iss-9/p22.html.

G. J. Whitrow's "Why Physical Space Has Three Dimensions" appeared in *The British Journal for the Philosophy of Science* 6:21 (1955): 13–31 and contains relevant historical information as well as the source for the arguments considered here. Stephen Hawking's argument, and a discussion of string theory, appears in chapter 10 of *A Brief History of Time: From the Big Bang to Black Holes* (Bantam, 1988).

Susskind's book is *The Cosmic Landscape* (Back Bay Books, 2006), and the argument that I discuss is found (mostly) in chapter 11.

See http://mathworld.wolfram.com/UniversalCellularAutomaton.html for a more technical review of the use of Life as a computer.

7

The Shape of Space II

Curved Space?

A typical school math curriculum includes the study of basic geometry in two and three dimensions (figure 7.1): for example, the Pythagorean theorem, that in a right-angled triangle 'the square of the hypotenuse is equal to the sum of the squares of the other two sides'; that the area inside any circle is equal to π times the square of the radius; the parallel postulate (that 'through any point there is exactly one line parallel to a given line'); and so on. These statements—and most likely anything else taught in precollege mathematics—all follow from the five laws (or better, the 'axioms') of geometry proposed by Euclid in the fourth century B.C. (Well, strictly speaking, there are some additional tacit assumptions.)

In this chapter we will investigate alternative, 'non-Euclidean', geometries: possible 'shapes' of space, in a sense different from that used before. These are geometries in which many of the things taught in school are false. However, we will start by thinking about the time before they were discovered (in the nineteenth century). Then—and likely when you learned it in school—Euclidean geometry was considered as being of a piece with all mathematical truths: as unshakable, indisputable, and certain as $1 + 1 = 2$.

7.1 MATHEMATICAL CERTAINTY

As such, mathematical truths pose an interesting problem: just what makes them so certain? Just why does it seem impossible that they might turn out to be false? There are, after all, lots of things that we're very sure of: for most people, who our parents are, that the sun will rise tomorrow, that we have two hands and, for you, that you are currently reading this book. But there are ways that we might be mistaken about any of these things: family secrets are sometimes revealed, the Earth could be struck by an asteroid which stopped it spinning tonight, perhaps one of my hands has been amputated and I simply experience a 'phantom' limb, and as the film *The Matrix* shows, I could be systematically mistaken about what I infer from my experiences: there may be no book at all. However, many

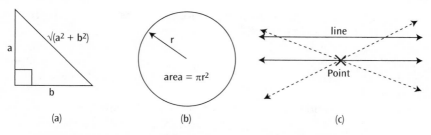

Figure 7.1 In Euclidean geometry, (a) the square of the hypotenuse is equal to the sum of the squares of the other two sides; (b) the area of a circle is $\pi \times$ the radius; and (c) there is a unique line through any point that is parallel to a given line (you can see that any other lines intersect the given line somewhere).

philosophers have found it hard to imagine that circumstances could ever show one to be wrong in thinking that $1 + 1 = 2$ or that the Pythagorean theorem is true.

The situation is particularly perplexing with regard to geometry, since it is clearly not just about mathematical entities (such as abstract numbers, points, lines, triangles, and circles), but also about *physical space*—about the points of the space we inhabit. It seems that Euclid's geometry tells us that parallel to the actual line along which a beam of light travels, there is only one other parallel to it through a given point; it also seems to tell us that a circle around the sun at the radius of the Earth's orbit has an area of π times that radius squared; and so on.

How can this be? Usually our knowledge of the physical world is not certain but is always open to revision by new experiences; what makes geometrical knowledge concerning space so special, so peculiarly certain? Well, we'll see it isn't at all, but that fact wasn't obvious before the nineteenth century, and I want to start our investigation before then. (One might ask whether the certainty of $1 + 1 = 2$ is any less perplexing. We'll leave aside the question of whether geometry is truly a special case in mathematics, but see the further readings.)

Immanuel Kant had an answer to this puzzle. He proposed that the geometry of space is not something *learned*, as are most facts about the world, but rather something *imposed*, involuntarily, by us on our experiences. His idea was that whenever we see, hear, feel, taste, or smell anything, our minds add considerable interpretative and organizational structure to our perceptions. As a consequence, we can never *directly* experience the world, since every experience contains an important, undeletable component that comes from us.

In particular, all our experiences are organized, according to Kant, within the framework of three-dimensional Euclidean space. Therefore we see things in such a space not because they are so arranged (they may or may not be), but because it is impossible to see things in any other way. Thus geometric truths about space get their certainty because they concern properties that we invariably impose on what we experience; no experience can disprove them. On the other hand, since geometry does concern our experiences of the world, it is clear that geometry concerns the world (as we experience it) not just abstract mathematics. (Kant believed that a similar story applies to other branches of mathematics, including arithmetic.)

You may be wondering how Kant thought that he knew that we always perceive things as if arranged in Euclidean space. Was it by reflection on his perceptions? Saying that would be saying that he discovered it by a kind of psychological experiment. But then Euclidean geometry would again be a matter of experience, and would not be certain; further, more careful psychological experiments could show that Euclid's axioms were not true of our experiences after all. Instead, Kant believed (mistakenly, we shall see) that he could prove that it followed from the very nature of experience that it was *impossible* to have experiences other than in Euclidean space: no Euclidean framework means, necessarily, no experiences. If he were right, then of course no experiment would ever show experience to be other than Euclidean; hence the claimed certainty of Euclidean geometry.

Kant's answer to the puzzle is clever and had a profound effect on philosophy in many ways. Not least, if correct, it would mean that in addition to the facts discovered by science, there is a class of 'metaphysical' truths that philosophy can discover: those concerning the structure that we impose on experience. However, developments in mathematics soon generated problems.

7.2 LIFE IN NON-EUCLIDEAN GEOMETRY

Soon after Kant wrote, it became clear, particularly in the work of Carl Friedrich Gauss (who never published his results, in part for fear that they were too radical), that other systems of geometry were as mathematically legitimate as Euclid's. In these geometries, those geometric facts taught in school may no longer be true at all: the square of the hypotenuse may not equal the sum of the squares of the other two sides; the area of a circle may *not* be a multiple of the radius squared, but some more complicated function; and there may be *many* lines through a single point that are parallel to a given line, or *none* at all.

It's not hard to see how this could happen in a two-dimensional space, such as a spherical surface. For instance, pretend that the Earth is a perfect sphere and consider a triangle drawn from the North Pole to the equator along the Greenwich meridian, then along the equator 90° east (from the coast of West Africa to close to Indonesia), and then back up to the pole (figure 7.2). This triangle is right-angled (in fact all of its angles are 90°), but each of its sides are the same length, 1/4 × circumference of the Earth (= about 6,225 miles), and so it doesn't satisfy the Pythagorean theorem.

Figure 7.2 In this right-angled triangle on the surface of a sphere, all the sides are equal, violating the Pythagorean theorem (in fact, all the angles are right angles, so all the sides are diagonals).

Maybe you are worried at this point. A triangle is a figure bounded by three straight lines, but on what basis do we say the figure just described has *straight* sides? What about the triangle formed by the lines between the corners that pass through the sphere? But remember, we are interested in the geometry of a two-dimensional space, the surface of the sphere, not the three-dimensional ball it encloses. The lines through the ball simply don't lie in the surface and so simply are not elements of the geometry under discussion.

But why are the lines we chose on the sphere the 'straight' lines in that space? Well, we all know that a straight line is the shortest path between any two points in Euclidean space, and that is a notion that can easily be carried over to other geometries. The lines we have drawn are 'straight' in the sense that they are the shortest paths between any two points that lie on them. As a matter of fact, they will therefore be arcs of 'great circles', circles that divide the sphere into equal hemispheres.

In any geometry, the shortest paths are the 'straight' lines; to avoid confusion with familiar Euclidean concepts, we shall refer to them by the technical term *geodesic* when we discuss non-Euclidean geometries.

Another kind of non-Euclidean geometry describes a two-dimensional surface in the shape of a saddle (figure 7.3): from the center it curves up to the pommel and cantle, and down to the sides (and off to infinity in all directions). Imagine a geodesic drawn on the saddle from the pommel through the center and up to the cantle, and a point off to the left. Because the surface opens up like the mouth of a trumpet to the front and back, geodesics are pulled apart, and there are many 'straight' lines through the point that don't intersect the given line (even when it and they are extended to infinity). These lines are *many* parallels to the given line.

If you think about it for a moment, it's clear that there are *no* parallel lines at all on the surface of a sphere: all the great circles intersect one another. There are of course smaller circles on the sphere, through a given point, that do not intersect a given great circle, but they are not

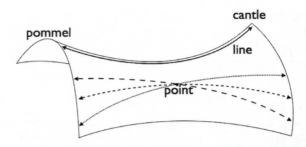

Figure 7.3 On the saddle, through a point there are lots of (actually, infinitely many) lines parallel to any given line—the three drawn here for instance.

geodesics and so they don't contradict the claim that no geodesics are parallel. Similarly, in the plane I can draw all kinds of curves through a point that don't intersect a given line; only one will be straight.

The plane and saddle have the same topological shape, which is different from that of the sphere, but in the everyday use of the term, we'd say all three geometries have different 'shapes'. Putting it another way, since the geometric properties of a space depend on how it is curved, geometry is the study of the 'shape' of space in the colloquial sense. In this sense we can say that the difference between Euclidean and non-Euclidean space is that between 'flat' and 'curved' space.

In general, two-dimensional non-Euclidean geometry is rather easy to visualize, but what about three (or more) dimensions. What does a three-dimensional sphere or saddle, for example, look like? We picture two-dimensional curvature by picturing the space from the outside, in three-dimensional space; to visualize three-dimensional curvature 'from the outside' we would have to picture the (three-dimensional) space in four or more dimensions, something I at least am unable to do.

I don't mean to suggest that a non-Euclidean space can exist, in some sense, only if it is embedded in a Euclidean space. Even a two-dimensional spherical space, which we can visualize as the surface of a sphere, could be the entirety of space in some possible world, not just a part of a larger 'real' three-dimensional space. What I want to point out is that our minds do seem constituted to picture things in three-dimensional Euclidean space, making it difficult to picture spaces that cannot be contained in three-dimensional Euclidean space.

To get around this difficulty, we will explore three-dimensional non-Euclidean geometry 'from the inside', considering how things would look to someone living in non-Euclidean space. To do so, we must make some assumptions about how things behave in space (assumptions we'll question later): let's assume that if you pull a rope taut—that is, pull it 'straight'—then it will follow the shortest path between its ends, a geodesic. Similarly, let's assume that light travels along the shortest and straightest path between two points. In particular, when you see something in the distance, light travels from it to your eyes along a geodesic. (This assumption is subtle in general relativity, in which light travels along non-Euclidean geodesics of space and time taken together. We will put those subtleties aside here.)

As in two dimensions, we can characterize the difference between Euclidean and non-Euclidean geometries by the Pythagorean theorem (and, as we'll see, areas of circles and the parallel postulate). In three-dimensional Euclidean geometry the square of the hypotenuse of any right-angled triangle equals the sum of the squares of the other two sides (it lies in a plane after all). In three-dimensional non-Euclidean geometries it will not; it lies in a non-Euclidean 'plane'. (By definition, we are talking about figures whose sides are geodesics—the shortest lines—between the corners. Think of the general 'plane' as the surface obtained from a point and a geodesic by drawing all the geodesics that cross both the point and line.) By extension we call Euclidean and non-Euclidean spaces of three (or any) dimensions 'flat' and 'curved', respectively.

So imagine the three following situations: (1) we pull cords tight between three corners to make a right-angled triangle; (2) we send laser pulses out at right angles to a pair of distant points, which then exchange a laser pulse; (3) we measure the distance to two distant stars seen at right angles to each other in the sky and then send a probe from one to the other along the direct line of sight (let's pretend that the stars don't move during this time). In each case we construct a right-angled triangle, either from rope, or from the paths of the laser pulses, or from the paths light travels between the distant stars and to us (since we sight along the path on which light travels to us, and the probe travels along the path of light). Given our suppositions about ropes and light (and a laser pulse is just light), their sides would all be geodesics.

If space is flat, we would find that the Pythagorean theorem holds: (1) the square of the length of the rope along the hypotenuse would be equal to the sum of the squares of the lengths of the other two pieces of rope; (2) the square of the time that the pulse takes to travel between the distant points would be equal to the sum of the squares of the time it takes to reach those points (light travels at the same speed between each point, so the time it takes a laser pulse to go between two points is proportional to the distance between them); and finally (3) the square of the distance that the probe measures between the two stars would equal the sum of the squares of their distances to us. However, if space were curved, then in each case we would find that the square of the length measured along the hypotenuse was not equal to the sum of the squares of the lengths measured along the other sides at all: the triangles marked out would violate the Pythagorean theorem.

You might think that there's nothing too odd about that: the same thing could happen if we didn't perform our experiments well. If the cord across the hypotenuse weren't tight enough, then our measurements would be inconsistent with the Pythagorean theorem; the same thing would happen if the laser pulse sent between the points passed through some medium (water, say) that made it travel more slowly than along the other sides; again, we would expect a violation of the Pythagorean theorem if we were careless about picking stars exactly at right angles. But if space is non-Euclidean, the violation of the Pythagorean theorem is not due to experimental error, but due to the shape of space itself. So pull as tight as you like on the rope, repeat the laser experiment in the vacuum, and find two stars that really are at right angles, and the violation remains; and use any other sensible method of picking out geodesics in space, and the same result obtains. That is, it is not the apparatus, but space itself that produces the result.

(Of course we could adjust the sides—by letting out the string, perhaps—to satisfy the theorem, but it's no surprise that we can find three bits of string such that the square of the length of one equals the sum of the lengths of the other two. However, once we let the strings go limp, they no longer mark out geodesics and so no longer follow the sides of a true triangle in space.)

The same holds for circles constructed in the 'plane': pull a cord tight from a center and mark out a circumference, or consider all the points reached from a center by a light shining in all directions after exactly one second or one year. If space is Euclidean, then the area is $\pi \times \text{radius}^2$; if the space is spherical, then the area is always more than $\pi \times \text{radius}^2$, by a factor that grows larger as the radius grows. Again, this result does not reflect any defect in the experiment.

Most counterintuitive perhaps, is the application of the parallel postulate. Imagine lining up people along a geodesic shoulder-to-shoulder with no end and no gaps—use light rays to line them up 'straight'—and imagine standing in front of them so that by turning your head you can see different people in the line. If you are initially looking at someone in line, turn your head 180° in the plane and you end up looking away from the line, seeing no one. Further, there is exactly one direction in the plane in which you can look both ways and see no one at all directly ahead, however far you look, namely along a line of sight parallel to the line of people. (Here we idealize that you are wearing blinkers so you have no peripheral vision, and that you have telescopes powerful enough to look as far as you like.) At least, these two things are true in Euclidean space; in non-Euclidean space, depending on the details of the geometry, these things may not be right at all.

For instance, in a spherical space you would see people in whatever direction you looked. The geodesics are great circles, which are closed; hence the line of people is closed, with you inside it! In this case there won't be any direction at all in which you fail to see someone in the line (figure 7.4(a)). As we said, there are no parallels on the sphere, so every geodesic—every path of light—to you passes through the line along which the people stand.

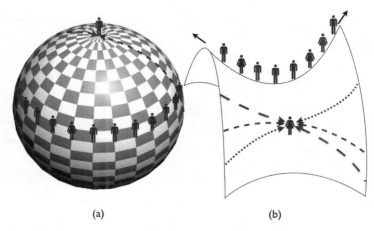

(a)   (b)

Figure 7.4 (a) There is a complete line (i.e., great circle) of people lined around the equator of the sphere. The person off the line sees people in either of the opposite directions shown; you can see the path of the light traveling along a geodesic (i.e., great circle) to him. In fact, whichever direction anyone looks, he will see some of the people, since every geodesic intersects the line. (b) In the saddle the person off the infinite line of people looking in any of the directions shown (and many others) will not see anyone in the line. The geodesics through her position do not intersect the line, and light reaching her travels along those geodesics.

In saddle geometry quite the contrary holds (figure 7.4(b)). There are many directions in which one can see nothing, however far in the distance, *and* turn around 180° and still see nothing, however far away. That is, as we saw earlier, there are many lines through a given point that are parallel to a given line, and so there are many parallel light rays in saddle space, not just one. (Again, this phenomenon is not due to a defect in the light reaching our eyes; we are supposing that the light is following geodesics, so the phenomenon is due to the geometry of space.)

7.3 WHAT KIND OF KNOWLEDGE IS GEOMETRY?

We started this chapter with Kant's account of geometrical knowledge, according to which our knowledge that geometry is Euclidean is certain. Clearly the discovery of non-Euclidean geometries has an important bearing on that view, but what exactly?

Let's unpack Kant's claim step by step and see what, if anything, of his idea is correct. The first point is that non-Euclidean geometries are just as logically consistent as Euclidean geometry, and so all geometries are on an equal mathematical footing, that is, when considered as descriptions of abstract mathematical points, lines, and surfaces. To ask which geometry is mathematically 'true' is like asking whether whole numbers *or* fractions are 'true'; they are both coherent systems and so are equally legitimate in their own domains. Similarly, it makes no sense to ask whether Euclidean or non-Euclidean geometry is 'true': each is correct in its domain.

So it doesn't make any sense to think that we can *know* which geometry is *the* unique mathematically true one, because there is no such thing. Similarly we can't know who was the captain of the Euro 2008–winning English soccer team; England was eliminated in the qualifiers so there was no such person. Hence, insofar as Kant claims that we can be certain that Euclidean geometry is the unique mathematical truth, the very possibility of non-Euclidean geometry—accepted only after his death—shows that he was wrong.

But remember, Kant's claim about our knowledge of geometry is more complicated than this response allows, because geometry is not just mathematics but also a 'theory of space'. The second part of his view is that we can be certain (perhaps as certain as that $1 + 1 = 2$) that the geometry of *physical space* is Euclidean. But surely that is wrong too: since the different geometries are logically possible, it should be logically possible for physical space to have any of the geometries, and so nothing logically necessitates space having one geometry rather than another.

Of course it doesn't follow that we can't be very sure about the geometry of space: nothing logically forces me to have the weight that I do at this moment, but by experiment we can become quite sure what I weigh (within normal experimental accuracy). It seems reasonable to believe that we can also use experiment to establish the geometry of space—not with mathematical or logical certainty, but as reliable empirical knowledge.

Indeed, the examples that we've already considered seem to support this idea. The experiments with the triangles, for instance, apparently constitute procedures for measuring whether space is flat or curved: if the measurements satisfy the Pythagorean theorem, then space is Euclidean, and if not, then it is not. That is, when we realize that Kant is wrong to claim that our knowledge of the geometry of space is certain, a natural response is to conclude that, after all, it must be experimental, like other knowledge of the physical world.

The experiments we discussed are of course idealizations, chosen for their simplicity, but it's important to realize that real experiments can be, have been, and are being performed to determine the geometry of space. Not surprisingly, the early pioneers of non-Euclidean geometry were interested in discovering the geometry of space, and they performed some tests, akin to ours (forming triangles with the Earth and a pair of distant stars at the corners), and found that within the accuracy of their experiments, space appeared flat.

Recent measurements of the heat left over from the big bang— the 'background microwave energy'—by the Wilkinson Microwave Anisotropy Probe (launched in 2001) showed that space is very flat indeed. (The inference to the geometry of space is less direct in this case than in our experiments.) However, as was discussed in chapter 4, the data does not rule out space having the shape of a 'closed dodecahedron', in which case it cannot be quite perfectly Euclidean. These experiments refer to the geometry of space over very large, intergalactic scales. The non-Euclidean geometry of space over shorter distances can be worked out from the arrangement of mass, using general relativity.

It certainly seems that we treat the geometry of space as an experimental matter, quite contrary to Kant's views. There is, however, one further component of his view, namely that however things actually are, we will necessarily *perceive* objects as arranged in Euclidean space. Is this, at least, true? Even if space is curved, and our experiments show that it is, might we still somehow have to experience them as if they were in Euclidean space?

We have to think carefully about what this claim means, given our discussion of non-Euclidean geometries. First, once you see their possibility, it is hard to see why no possible sentient being could experience them for what they are. We'll return to this question in the next chapter. Even so, it is still possible that humans, because of the way our minds work, can experience space only as Euclidean. That would be a matter for cognitive experimentation, contrary to Kant's view, but if true, an attenuated version of his claim. Again, we will discuss that possibility in the next chapter.

Second, it's not so clear what it means to 'experience space as Euclidean'. On the one hand, the very experiments that we have been describing would, if they showed violations of the Pythagorean theorem and parallel postulate, amount to experiences of non-Euclidean space. After all, we would see the measurements being carried out, and the observed length of the hypotenuse squared is not equal to the sum of the square of the observed lengths of the other two sides.

But then again, what if triangles that violated the Pythagorean theorem *appeared* to have curved sides? We might measure them to be straight— see that string lay along them when pulled tight, and that light traveled along them—and so might infer a non-Euclidean geometry, but still they would *look* curved to us. Or again, if we found there were many directions in which we could not see people, from which we infer a saddle geometry space, the line of people might just look curved. Wouldn't that be a sense in which space is experienced as Euclidean? And if our minds are built so that we will always see things thus, wouldn't there yet be something right about what Kant says?

What we need to address these issues, then, is some account of what it is to experience space as having one geometry rather than another, and some empirical criteria that will tell us how a subject is experiencing space. Then we can perform experiments to test how space is perceived under various stimuli. Again, that issue will be taken up in the next chapter.

The question addressed in this chapter has been how mathematical knowledge—which seems to be certain—relates to experimental knowledge of the world, especially our knowledge of space. According to Kant, experiences are necessarily of things in Euclidean space, regardless of how things actually are, hence the certainty of Euclidean geometry and its relevance to experimental matters. But the existence of non-Euclidean geometries undermines his view: no one geometry can be said to be uniquely correct in the way Kant supposed. So it seems that geometry is an experimental matter.

However, in the next chapter we will see that things are not quite so straightforward; we will see how the relationship between mathematical and physical geometry is rather more complex; perhaps geometry is not experimental either.

For now, we have seen again how advances in mathematics and physics bear on important philosophical questions. That lesson was not so surprising when we were considering questions about the shape of space, since physics studies space. In this chapter the philosophical issue concerned, not some physical entity, but the nature of knowledge, the subject of 'epistemology'. We have seen that even this topic is far from independent of mathematical and physical discoveries.

Further Readings

The home page for the Wilkinson Microwave Anisotropy Probe is http://map.gsfc.nasa.gov. It contains a lot of very accessible material on the probe and what it is showing us about the geometry and topology of space.

This chapter invokes an intuitive idea of mathematical certainty. Stewart Shapiro's *Thinking about Mathematics* (Oxford University Press, 2000) contains a more complete, but still introductory, discussion of this question, as well as other issues in the philosophy of mathematics.

A nice further discussion of non-Euclidean geometry can be found in Rudolf Rucker's *Geometry, Relativity and the Fourth Dimension* (Dover, 1977).

8

Looking for Geometry

The last chapter suggested that it would perfectly possible to measure the geometry of space: pull cords between three points to make a right-angled triangle, and measure the sides; or time laser pulses between the same three points; or send a probe between two stars at right angles to the Earth. If the hypotenuse squared equals the sum of the squares of the other two sides... well then the geometry of space is Euclidean. If not, then space is non-Euclidean.

However, Henri Poincaré, an important figure in the development of non-Euclidean geometry, revealed an important problem with drawing this kind of conclusion.

8.1 MEASURING THE GEOMETRY OF SPACE?

Suppose we found that the triangle formed from the three ropes satisfies the Pythagorean theorem. Assuming that the ropes follow straight lines in space, we infer that space is Euclidean; however, what justifies the assumption? If we assumed that space is Euclidean, then we could infer from the experiment that the ropes followed straight lines, but that would be to reason circularly. To say space is Euclidean because ropes are straight because space is Euclidean is just to say we think space is Euclidean because we think space is Euclidean, that is, to give no independent reason at all.

We could instead assume that space was saddle-shaped, in which case the experiment would show that the ropes were following curved lines. In fact, given that assumption, we could infer from such experiments a new law of statics: that ropes under tension *do not* tend to straight lines, that they do not represent geodesics.

Or suppose we find that there are a range of angles in which we can look and not see the line of people in either direction. If we assume that light follows geodesics, then we infer that space is non-Euclidean. But what justifies that assumption? If instead we start with the assumption that space is Euclidean, we can instead infer that light travels along *curved*, not straight, lines; our line of sight curves away from the line of people. Either way the observations are the same. And so on.

Poincaré claimed that *any* such experiment could be interpreted in terms of both Euclidean and non-Euclidean geometry: whatever the outcome, assume whichever geometry you like and draw the appropriate conclusions about the relations between physical objects and geometric figures, whether the objects are straight or not, for instance. (More precisely, he claimed that any observations consistent with Euclidean geometry would also be consistent with saddle geometry, and vice versa.) He concluded that no experiment could refute your assumption and hence that the geometry of space was not something that could be settled experimentally. But that leaves him with a puzzle: if the geometry of space is a matter neither of mathematical certainty (because of the multiplicity of geometries) nor of experiment, then why do we seem to know it?

His solution was to suggest that the different interpretations of the experiments—Euclidean geometry and straight ropes versus non-Euclidean geometry and curved ropes, say—do not correspond to different factual descriptions of the world. How could that be? How could there be no real difference between space being flat or curved? Or ropes being straight or wavy? Even if it is impossible to tell the difference, isn't there a difference? Poincaré proposed that the 'disagreement' was just like that between people using different units of measurement.

Imagine two surveyors, one working in feet and one in meters. Suppose they measure the height of the Sears Tower; the first concludes that it is 1,450 feet tall while the other demurs; 'no, it's 442 meters tall'! Of course there is no factual disagreement about the height, only about how to quantify it.

Or again, as I write, the kilogram is defined by a platinum-iridium block in a vault in Paris, but in the next few years that standard is likely to be replaced by an 'experimental' definition. Perhaps the kilogram will be defined as the mass of a fixed number of atoms of gold or silicon. Why would it? First because such a definition would allow scientists to calibrate equipment without having to compare their instruments to the standard in Paris (via a series of intermediary weights, since the original itself is rarely disturbed). Thus the new definition is worth adopting only if there is a practical way of counting atoms; because this is not an easy matter, such a definition has not been adopted before now, and another definition may yet prevail.

Second, the definition might be changed to improve accuracy; a metal body never stays the same mass, because atoms evaporate off it, so the kilogram changes slowly over time. This is not a hypothetical issue; when the standard was placed in its vault in 1889, duplicates were stored with it, but they now have measurably different masses. The new definition would avoid this problem and make the kilogram more accurate, though this is not to say that scientists will ever have a perfect kilogram. There are inevitable errors in any actual experimental calibration: perhaps there are ten extra atoms in the sample, or perhaps some of them are excited and so have extra mass, and so on. The point is to make units as precise as they need to be for the measurements science needs to make. (The meter, for instance, used to be defined by a metal bar, but for similar reasons is now defined as 1/299,792,458 of the distance traveled by light in one second.)

So if our first scientist employs the old definition of the kilogram and the second the new one, when they weigh objects they will again have disagreements. Especially, the latter will be able to measure meaningfully to more decimal places. But as before, this is not a factual disagreement about the object being weighed, but only about definition of the units to be used. The disagreements in both cases can be traced to the different free choices—'conventions', to use Poincaré's term—that the scientists adopt; free because neither mathematics nor experiment forces one choice over another. But of course some choice must be made if we want to use a quantitative measurement scale at all.

Poincaré suggested that the disagreements over the interpretation of geometric experiments should be understood in just this way. It is immensely convenient to describe physical processes and systems in physical space, but if we want to do so we first have to make some choice—a convention—about the geometry of space, just another, more complex, 'choice of units'. Then the different interpretations of geometric experiments arise from different choices about geometric 'units'; in one choice, tight ropes and light rays are straight, in another both are curved. But these disagreements are no more about spatial facts than our scientists' disagreements concerned facts about lengths or weights. Poincaré thus placed geometrical knowledge between mathematical certainty and experimental fact—a free convention required before experimental facts could be interpreted.

The proposed analogy between geometry and units is easier to understand if we think about what our geometric experiments actually establish. After all, they have different possible outcomes—either the measurements satisfy the Pythagorean theorem or they don't—so what is being measured if not the geometry of space? Poincaré's answer is that one discovers laws that govern the way bodies move relative to one another, not anything about the geometry of space or the relation of bodies to it. The key idea is that of a mathematical 'group', and not only here: the group is one of the most important concepts of modern physics.

Once again, it is fairly simple to understand this idea in two dimensions, and though it is rather harder to visualize in three dimensions, the meaning is the same. So consider the (Euclidean) plane and the (non-Euclidean) surface of a sphere the size of the Earth, and imagine a two-dimensional stick figure in each; on the sphere we'll have him at the equator, oriented north-south along the Greenwich meridian. In both cases there are the same kinds of possible motions: we can slide (or 'translate') the figure around the surface along any line, or we can rotate it

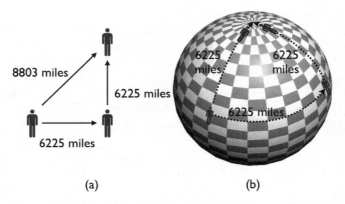

(a)   (b)

Figure 8.1 (a) In the plane, moving to the right then up to a point has the same effect as moving there directly. (b) But in the geometry of the sphere, the effects are quite different; for instance, the traveler here ends up facing in a different direction.

about any point. But what distinguishes the spaces is which combinations of motions produce *the same change*.

For instance in the plane (figure 8.1), sliding our stick figure 6,225 miles to the right then 6,225 miles up leaves him just the same as if he were simply translated 8,803 miles along a line at 45° (the paths are the sides of a right-angled triangle, so 8,803 comes from the Pythagorean theorem). But in the surface of the sphere, translating the stick figure 6,225 miles to the east (1/4 of the way around the equator) and then 6,225 miles to the north leaves him at the pole, facing away from Greenwich, while translating him directly to the pole leaves him facing 90° from Greenwich. (And a straight-line motion at 45° for 8,803 miles won't leave him at the pole at all.)

Generally, in Euclidean space, motion along two sides of a triangle has the same result as motion along the third, but on the sphere it does not. Similarly, if the stick figure moves 24,900 miles in any direction around the sphere, the effect is the same as if he did not move at all, but the same motion in the plane leaves him somewhere else. And if we considered the same motions on the saddle, we would find different results again.

The 'group of displacements' is simply the set of rules that determine which sequences of motion are equal to each other. As the example shows, different geometries have different rules: two sequences of motions that have the same effect in one geometry may not in another. In fact, any two geometries will have different groups of transformations, and so we can uniquely characterize geometries by their groups, and this is what Poincaré proposes for space.

8.2 THE 'GEOMETRY' OF POINCARÉ'S SPACE

'But wait', you should be saying. 'How does this help with our problem? Didn't we see that we can pick any geometry we want to describe space, as long as we accept the appropriate rules for the behavior of bodies. So since each geometry determines a different group, it follows that we can pick different groups to describe space, as long as we accept the appropriate rules for the behavior of bodies.'

Poincaré's response is that while a geometry describes the shape of space itself, a group of transformations can be taken to be a description of the motions of bodies in space directly, which captures everything that we can hope to discover in our geometrical experiments. That is, our experiments establish which actual paths have the same effect on actual physical bodies, something captured by the group. And those effects—and hence that group—are independent of how we interpret them in terms of geometry and the behavior of bodies.

Poincaré gave a philosophical fable that helps explain this point as well as the others we have been discussing. Far away and long ago there was a three-dimensional Euclidean universe with the (topological) shape of a ball—space as Aristotle conceived it. In this space everything shrank to nothing as it moved from the center to the edge (as before, by a factor of $1 - w^2$, where w is the fraction of the distance from center to circumference; see figure 4.2). Also in this space, light was bent and slowed down according to suitable rules (given by Poincaré), again depending on the distance from the center. Since everything got smaller at the same rate, the inhabitants found it impossible to use anything to measure the shrinkage—any conceivable measuring object, including their own bodies, was shrunk to just the same extent—and so the contraction was quite undetectable.

Now one of the effects of the shrinkage was that whenever the inhabitants of the world measured the lengths of various paths using rulers or cyclometers or whatever to find the shortest one, they found that it was the same as the bent, non-geodesic path traveled by light. (Why is explained qualitatively in figure 8.2a.) Similarly, when they pulled a rope tight, they found that it lay along the same shortest measured path between its two ends. Since the rope also contracts, it too is shortest when it lies along that path.

And so, not only were they unable to detect the shrinkage of objects—since it happened uniformly to all of them—they were also unable to detect the bending of light, since every way they had of figuring out which paths were geodesics, with light or rope or by measuring the shortest paths, gave the same answer.

Of course, since the triangles they constructed with light and rope were curved, they found that the Pythagorean theorem was violated. And since light curved for them, they found that they could look both ways

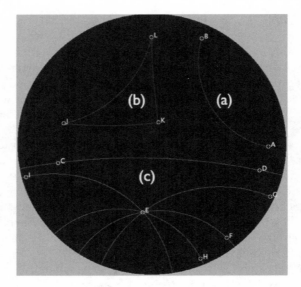

Figure 8.2 In Poincaré's space, all the lines shown are measured to be straight: light travels along them, ropes pulled tight between points on them lie along them, and when their lengths are measured they are found to be the shortest lines between any two points on them. (a) This last point can be seen by considering AB. One way to measure the length of a curve is to lay identical rulers along it; one curve is shorter than another if fewer rulers can be lined up between the ends. Although AB is longer than a straight Euclidean line between A and B, it is measured to be shorter because in Poincaré's space the rulers used to measure it are larger because they lie nearer to the center. (b) The three sides of JKL are measured to be the shortest lines between the corners; they are the lines that light travels, and tight ropes follow. There is a right angle at K, but $JL^2 \neq JK^2 + LK^2$, violating the Pythagorean theorem. (Here, for instance, JL means the measured length of the line between J and L.) (c) Finally, CD is an infinite line and E a point not on it; all the other lines are examples of lines that do not intersect it. But all of them are measured to be straight, so measurements reveal infinitely many parallels. For instance, if there were people lined up all along CD from end to end, then if one were to stand at E, none of the people could be seen in the direction of any of the lines. That is, the lines all show possible paths of light.

along many paths and fail to see a line of people; there were many paths along which light would reach them from either direction that didn't come from somewhere on the line. And so every experiment they ever performed had exactly the same result as if they had performed it in non-Euclidean saddle space with rigid objects and light traveling along geodesics.

In particular, experiments to see whether identical and adjacent objects are again identical and adjacent after traveling different paths showed that the displacements of physical bodies obeyed the rules of the group of geometric displacements in saddle space (the 'saddle group').

According to Poincaré's philosophy, there is in fact no geometry to these people's space, only a series of experiments involving comparisons of objects as they move around, whose outcomes can be described in both Euclidean or non-Euclidean terms as we choose, just as measurements can be described in feet or meters according to one's convention. But the outcome of the experiments is not ambiguous: in particular, that the displacements of physical bodies obey the rules of the saddle group is not something that is conventional. (*Why* they do is conventional: because space is flat and bodies change size as they move, or because space is non-Euclidean and they don't?)

Knowing the group of physical displacements is an extremely powerful piece of information. If it is the saddle group, then bodies move *as if* they were rigid geometric figures in saddle space, which tells us the results of any of our (or any other) geometric experiments. Experiments will violate the Pythagorean theorem, for example. If we find that the group is Euclidean, then we know that bodies move *as if* they were rigid figures in Euclidean space, and hence know what will happen in geometric experiments. But of course in Poincaré's view it is only 'as if': while the group of displacements of physical bodies is extremely informative, it is a complete mistake to think that it is information about the geometry of space: it is simply information about how bodies move in relation to each other.

Now, according to Poincaré, experiment does not determine geometry, so we have a free choice from which we pick. But because bodies move according to some group or other, the choice is not between equals. Just as we have a free choice between different definitions of the kilogram, it does not follow that all are equally good; indeed, we saw reasons for making a change. Similarly, while we are free to pick any geometry for space, not all choices make equal sense; the most natural one, the one to which our thought is most attuned by constant experience, is the one corresponding to the group of physical displacements. In our everyday experiences, bodies move according to the Euclidean group, so while we could choose saddle space geometry for space, it would be extremely inconvenient to do so.

However, quite the opposite is true for the inhabitants of Poincaré's sphere; since bodies there move according to the saddle group, they will find it most convenient to adopt saddle geometry as their convention, though they would make no factual error to do otherwise but would simply make life hard for themselves.

8.3 HOW TO DISPROVE A DEFINITION

Taking a step back, Poincaré has identified what he takes to be the basic facts about space, namely the group of the displacements of bodies, and asked how these facts are captured by physical theory. What we see is that such a theory captures them by making them part of a theory that seems to imply extra facts: not just that two paths have the same effect on bodies, but that the space they lie in is Euclidean. But he compares this 'extra' to the fact that, say, this ruler is not just twice as long as that, but they are 2 m and 1 m, respectively. We realize that these latter are not really additional facts but describe the relative lengths of things, in terms of the definitions of units that we have made: at one time, in terms of the lengths of the two rulers relative to the standard meter. Choosing these units gives us a useful way to describe the relative lengths, but there are other ways to do so—in feet, for instance—that differ only in convenience.

According to Poincaré, though, the definition of length requires not only a conventional standard, but also a conventional geometry. In his world, if we pick Euclidean geometry, then we will say that the rulers are now both 1m long, if the shorter stays at the center and the longer moves $1/\sqrt{2}$ of the way from the edge (since $1 - (1/\sqrt{2})^2 = 1/2$). If we pick saddle geometry, then we will say that they are 2 m and 1 m still.

These claims differ not only in the units used, but in the relative lengths ascribed; according to one, the rulers are the same length; according to the other, one is twice as long as the other. Such claims sound factual, but they are supposed to be no more different than if we were comparing feet and meters; the differences are verbal. The choices of a standard and a geometry give us a useful way to talk about displacements, but there are other ways to talk that also differ only in usefulness and not in factual matters.

Poincaré's point is to restore something of Kant's view, against the idea that physical geometry is simply experimental. The geometry of space is 'certain', as Kant held, but now in the sense that we choose it, prior to any experiments on the behavior of bodies. Thus experiments can teach us nothing about geometry, only about the behaviors of bodies. (Of course, Poincaré's account of the certainty is quite different from Kant's; the latter thought that Euclidean geometry was necessary for any experience, not something about which we could make up our minds.) Poincaré illustrates how defining concepts is essential to science, and especially how difficult it can be to see where definitions are being made. There are certain things that we take as given in order to make other claims meaningful, but just how we do this is not always transparent. The meter is defined quite explicitly, but there may be other definitions that are tacit.

There are certainly ways to disagree with Poincaré. First, one might dispute whether the group captures all the facts about space. Most contemporary philosophers would not accept the idea that just because we can't observe something directly—a unique geometry—it follows that there is no such thing. For instance, we think we know a lot about subatomic particles, even though our evidence is indirect. But there are special reasons to be skeptical in the case of space. In the next chapter we will question what space is, and perhaps we will conclude that it is somehow unreal, not really the kind of thing that can have a particular geometry.

Another objection is that Poincaré overestimates the resilience of definitions against new experimental knowledge. We've already seen that definitions of lengths and weights change, as Poincaré was well aware, so why not geometry? If nineteenth-century experiments had revealed the results of Poincaré's ball, would physicists really have stuck to Euclidean geometry? It seems more likely that they would have taken space to have a saddle geometry instead. Of course, that wouldn't show that they thought geometry a factual matter to be decided by experiment; instead one might stick with Poincaré's view and conclude that they had revised their conventions. In other words, one would take the view that experiment can revise our opinions of which conventional geometry is the simplest to adopt. That is, if all bodies change size as they change place according to your convention, perhaps it would be more convenient to pick a new convention in which they don't!

An even more radical change is conceivable: one in which the definition turns out to be inapplicable because experiment leads to a theory in which geometry cannot be freely chosen. We don't have to imagine what such a theory would be like, because that is exactly the case in general relativity. As was discussed in chapter 1, in general relativity, the geometry of space—or more fundamentally the geometry of space*time*—depends on the arrangement of matter and vice versa. The point is that space is not at all a passive container in which things are located, but rather it reacts to the arrangement of matter; according to Einstein's laws, space reacts to matter just as much as matter reacts to matter.

Hence only certain combinations of geometry and distributions of matter over time are possible; in other conceivable combinations space is not reacting in the right way. In practice, out of a collection of combinations indistinguishable by experiment according to Poincaré, there's no reason to think that more than one is actually possible according to general relativity. In that case, given general relativity, experiments are unambiguous in the sense that only one geometry is compatible with a given arrangement of matter.

This is not quite to say that it would be impossible to first define the geometry of space and then insist that the rest of physics respect that geometry. Even keeping general relativity, one could insist on a particular geometry and posit new kinds of matter and energy so that, taken with the familiar kinds, they balance Einstein's equations. The new kinds would have to have no direct effect on matter and energy presently known to science to explain why they are as yet unobserved. (In fact, for rather different reasons, physicists do make such a posit: you may have heard of 'dark matter', which interacts with space but not with light or atomic matter.) What general relativity does show is that Poincaré was wrong to think that physics *had* to proceed by taking geometry as definitional, since the theory does not work that way.

Moreover, we've learned that while it's true that science treats some basic ideas as being true by definition (and maybe geometry used to be among them), such definitions still depend on how the world is, and so they may be revised as we learn more. For example, Aristotle defined 'down' to be the center of the universe, but now we know there is no such place, so that definition fails to pick anywhere out. Now we see that general relativity (given only familiar matter) means that we cannot choose a geometry by definition. In both cases, scientific advances 'disprove' definitional claims.

(General relativity has another very important consequence. A displacement group uniquely characterizes a geometry *only if the geometry is the same everywhere in space*. But in general relativity the geometry can vary from place to place, so no group captures all the facts about how bodies compare after taking different paths.)

One final remark. Indistinguishable geometries reemerge in string theory in the form of 'dualities'. There is theoretical evidence that the experimental behavior of strings is indifferent to whether cylindrical dimensions are very small (say 10^{-85} m) or very large (say 10^{10} lightyears, the size of the visible universe). String theorists seem to take the same line as Poincaré, saying that the choice of circumference is conventional; see Brian Greene's book in the Further Readings.

8.4 EXPERIENCING SPACE

While general relativity is relevant to our scientific knowledge of space, it doesn't help justify or explain our everyday understanding of geometry, since we have that quite independently of any knowledge of general relativity. People implicitly know the geometry of the world even if they've never heard of the theory, as of course they did before Einstein discovered it. For instance, judging the trajectory of a projectile, or figuring out the direction of the shortest path from one place to another, makes implicit use of geometry. When we figure out where a ball will land or the best route between a series of points, we instinctively make the assumption that Euclidean geometry holds (which it does essentially perfectly).

But this practical knowledge cannot be arrived at using general relativity, because the kinds of experience that lead us to it—watching projectiles, moving from point to point—themselves do not support general relativity. The observations that give credence to that theory are obtained through complex experiments, well outside the realm of everyday experience. And thus, while general relativity shows that Poincaré is wrong about our *scientific* knowledge of geometry, he may still be right about our everyday, *prescientific* knowledge. If our everyday experiences don't support general relativity, then we can't assume general relativity when we ask what can be inferred from everyday experience; hence all the different interpretations in terms of different geometries are apparently equally well supported by everyday experience. In the context of everyday knowledge, Poincaré's problem remains.

So what follows if Poincaré is right, if not about our scientific knowledge of geometry, at least about our prescientific knowledge? That is, suppose that what we know about geometry, before we learn about general relativity, is that bodies move according to a particular group: in particular for us, the Euclidean group. First we can, with Poincaré, criticize the final, mildest consequence of Kant's view: that humans are so constituted that they can *experience* the world only as Euclidean.

This time Poincaré does think that there is a real difference between Euclidean and non-Euclidean experiences, but he claims that they are equally possible for humans. As his fable illustrated, humans are just as capable of experiencing objects moving according to a non-Euclidean group as according to the Euclidean group (if that was how they actually behaved): for instance, nothing would prevent my seeing that motion along two sides of a triangle was not the same as motion along the third, or that a motion in a straight line could have the same effect as not moving at all. Therefore, if the group of displacements captures our prescientific knowledge of geometry, then we certainly could experience non-Euclidean geometries if we experienced bodies moving according to a different group; and if we did, such experiences would be different from the experience of Euclidean geometry.

It's clearly part of Poincaré's account that geometry—the group that is—is learned, not innate, which is why, contrary to Kant, that it would be possible for humans to experience space as non-Euclidean. This claim is pretty interesting, because it is apparently a testable hypothesis about the human mind. Is our (intuitive) belief that space is Euclidean—that objects move according to the Euclidean group—hard-wired, a matter of 'nature', or is it learned, depending on 'nurture'?

Cognitive scientists have paid considerable attention to this kind of question, though the answers are still a matter of contention (as is its assumption that we do in fact have an intuitive belief that space is Euclidean). However, a widespread current view, based on an explosion in research on the development of thinking skills in babies, is that our spatial knowledge is innate. For example, very young infants show little understanding of the three-dimensionality of space, but at about five or six months they rapidly develop the ability to discern differences in depth (for instance, a tendency to look more at '3-D' images on a screen or to reach for objects that look closer but really are not). But it is at this age that the visual cortex develops, and so these new spatial abilities are thought to be caused by the growth of specific parts of the brain rather than by experience.

In the context of Poincaré's specific proposal—that our geometric knowledge is learned knowledge of the *group* of transformations—the work of cognitive scientist Roger Shepard is especially interesting. He proposes that our reasoning about the motions of unseen objects— imagine a bird flying behind a stand of trees and reappearing, and figuring out if it was the same bird—relies on the use of the Euclidean group, which seems to support Poincaré's account of our perception of space. Moreover, Shepard argues that the use of the group is innate, developed by evolution, which is a modification of Kant's position.

8.5 WHERE IS GEOMETRY?

The last two chapters have led us through some pretty deep waters. The issue is that the shape of space seems to lie at the bottom of physical inquiries: if our fundamental picture of things is largely geometrical, then it's not surprising that the geometry of space is going to be a basic issue.

The discovery of non-Euclidean geometry refuted Kant's idea that geometry is necessarily Euclidean: mathematically each geometry is equally legitimate, and it seems an open question which correctly describes space. Kant also claimed that experiences are necessarily of things pictured in Euclidean space, that geometry is in the mind of any being capable of perception. Given the existence of non-Euclidean spaces, this claim seems unlikely at best; why shouldn't creatures in a saddle space (or Poincaré's space) experience things as being in saddle space? On the other hand, we've just seen that there are experimental reasons to think that Euclidean space is in some sense innate to humans (because of evolutionary pressures rather than because of the 'nature' of experience). There may be *some* geometry in the mind, but not all of it.

In the face of the failure of Kant's program for understanding the geometry of space, it's natural to conclude that it is an experimental matter: physical geometry seems to be 'out there' in the world. Poincaré of course resisted that idea, making geometry a matter of definition— a conventional 'choice of units'—for describing the fundamental spatial facts, contained in the group of displacements of bodies. For Poincaré, geometry is in the language in which we choose to talk about the world.

Really there are two central issues that Poincaré addresses: the question of the most basic spatial facts, and the question of the role of definitions in science.

First, Poincaré proposes that spatial properties belong not really to space itself but to *bodies* (or, if we are going to be more accurate, to our *experiences* of bodies). In his view, there are facts about what happens when pairs of identical bodies follow pairs of different paths to the same place; that information constitutes the group of displacements. For him, any time we seem to ascribe properties to space itself, we are just utilizing a convenient, powerful way of talking about the group.

In chapter 9 we shall consider a similar but distinct proposal; that the basic spatial properties are the distances between bodies. Again on this view, when we say that space has this or that property, we should not be taken literally but considered as using a useful system for talking about bodies. And why might we hold either this new view or Poincaré's? In the first place there is his argument that what we learn from experience are the spatial properties of bodies; so what might lead us to infer in addition that space itself has geometric properties? Or again, there is the radical intangibility of space. We can see and feel and manipulate material bodies; they are the basic furniture of the world. But space itself is far less substantial and is arrived at indirectly from our experiences of bodies; perhaps there are only the bodies?

The second component of Poincaré's proposal is his account of conventions in science. Of course science has to take some concepts in order to pursue experimental investigations: for instance, it took suitable definitions of time units and positions and motions relative to the fixed stars to produce meaningful astronomical data. But Poincaré's point is that such conventions may occur in places we do not expect, explaining why some scientific claims seem oddly resistant to experiment: they are conventions too.

Really his analysis of geometry is a beautiful example of how physics and philosophy can be integrated. His fable demonstrates the problem of knowing the geometry of space, he utilizes the mathematics of the group to capture the fundamental spatial facts, and finally he argues that the gap between the facts and our 'knowledge' is filled with a definition. In other words, he brings together the philosophy of knowledge, mathematics, and the philosophy of language to explain the logic of geometric knowledge. Of course, his analysis is not supposed to be of philosophical interest only, classifying different kinds of knowledge, but to give physicists a clearer picture of the terrain in which they work—a view that will prevent them from trying to test definitions, for instance!

Of course, we have seen that Poincaré underestimated the flexibility of scientific theories and their ability to change in the face of new evidence. While his account may have been accurate for prerelativistic physics (though even this idea is disputed), general relativity so drastically changed the landscape that geometry is no longer treated as a matter of definition. Still, that conclusion does not undermine the underlying point that physics needs an awareness of just what its concepts mean: in particular, how they are used within a theory. Poincaré undertook this philosophical task of trying to understand how the multiple new geometries should be understood in physics; however, his analysis had a profound influence on the development of physics. It revealed an 'equivalence' between forces (those shrinking bodies) and geometry–the same kind of equivalence that equates gravity and geometry in general relativity. Again, philosophical considerations are at the heart of advances in physics.

Further Readings

I recommend chapter 1 of Wesley Salmon's *Space, Time, and Motion* (University of Minnesota, 1981) for an investigation of many of the issues in this chapter, though it has a rather different take on Poincaré.

Poincaré's fable and arguments are found in his fantastic book *Science and Hypothesis* (1903), recently republished in *The Value of Science: Essential Writings of Henri Poincaré* (Modern Library, 2001). My discussion of Poincaré largely concerns chapter 5 of his book, with some reference to chapters 3 and 4. I also recommend Peter Galison's *Einstein's Clocks, Poincaré's Maps: Empires of Time* (Norton, 2004) for the discussion of Poincaré's involvement with standard units. I don't particularly like his treatment of conventionality, however; I have drawn on the more technical *Understanding Spacetime* by Robert DiSalle (Cambridge University Press, 2006).

The International System of Units is defined and discussed on the website of the National Institute of Standards and Technology at http://physics.nist.gov/cuu/Units/units.html.

Figure 8.2 was drawn using the wonderful applet *NonEuclid*, created by Joel Castellanos and available at http://cs.unm.edu/ joel/NonEuclid/ NonEuclid.html. It's very intuitive to use, and very helpful for visualizing Poincaré's space.

Brian Greene discusses spacetime 'dualities' in his book on string theory, *The Elegant Universe* (Vintage, 1999).

Finally, Jacques Mehler's *What Infants Know: The New Cognitive Science of Early Development* (Blackwell, 1993) is a good starting point for a popular introduction to work on child development, relevant to the discussion of the innateness of spatial knowledge.

9

What Is Space?

We've been discussing space without paying much heed to the question of what exactly it is we're talking about. And we've managed to do that because we've either talked about spaces in the mathematical sense, or referred the topological and geometrical properties of physical space to the behaviors of tangible things such as bodies and light. But what is physical space itself? Is it (as we started to suggest at the end of the previous chapter) nothing more than the spatial properties of tangible things?

I will end up proposing such a view, and elaborating it in later chapters, but it is by no means uncontroversial. For one thing, our language suggests otherwise. The furniture in my room 'occupies' some of its space, suggesting that there are two things where my chair is: the chair *and* the space it fills. An army cannot occupy a territory unless the territory exists, and linguistic analogy suggests that a chair cannot occupy a region of space unless that region exists either. That is, if we take such language literally, then there are *two* different things located wherever a body is: the body itself *and* the region of space it fills; my chair and a chair-shaped region of space are *coincident*.

Well, language can be misleading—recall Aristotle's argument for three dimensions—so we can't be completely satisfied with such reasoning. Moreover, space is a rather odd kind of thing, surely part of the physical world and yet somehow less substantial than tables, chairs, or even liquids, gases, atoms, or subatomic particles: material things cannot penetrate solid matter but bodies can penetrate space, since they occupy regions of it; space is somehow nothing but emptiness or something that is nothing; matter acts on matter (e.g., billiard balls bounce off each other), but can something as intangible as space do anything to matter (or vice versa)? Space's lack of tangibility suggests that it is nothing real after all.

However, if we don't take our talk about space literally, we should try to understand what it does mean. Such attempts will be discussed in this chapter.

Linguistic considerations may help illustrate how we think about space, but they are not the real reason for asking what space is. The question is important because understanding what motion is requires considering what space is. Not surprisingly, since motion is change with respect to space, different accounts of space will lead to different conceptions of motion. But 'motion' is a precise technical term in physics, for the laws of physics describe the motion of bodies. Thus only the right conception of motion—and hence of space—will make sense of the laws.

It was precisely the issue of coming up with the right conception that motivated Descartes and Newton (and others) to investigate the notions of space and motion: they were attempting to explicate the concept used in their theories of motion. Once again, we will see how apparently philosophical considerations were central to the founding of a new science.

9.1 SPACE=MATTER

The situation in which two very different kinds of things seem intimately related in a mysterious way is quite common in philosophy, and so we can learn a lot by thinking about the kinds of responses one can make in the case of space and matter; analogous problems have analogous possible solutions. (For instance, the responses to the space–body problem that we'll discuss correspond to different views about the 'mind–body' problem.) One such response is to claim that space and matter are literally one and the same thing. This remarkable idea has an impressive pedigree: it was held by Plato (in his *Timaeus*) and Descartes, was considered seriously by Newton, and is intimated by certain contemporary theories of space.

We can classify its proponents into two camps according to how they answer the following question: *Is the motion of a piece of matter always accompanied by the motion of an equal piece of space?* If not, then the view gives primacy to space and sees matter as a property of space; the motion of matter involves a change in the properties of space, not the motion of space. Newton discussed such a view, suggesting that we might understand matter in terms of certain regions of space having the property of impenetrability. Then the movement of bodies is simply a rearrangement of which regions are impenetrable: my walking across the room is nothing but a continuous change in which Nick-shaped region of space is impenetrable.

The development of non-Euclidean geometry allowed another negative answer to the question: that matter is a geometric property of space. For instance, William Kingdon Clifford proposed in 1876 that any material object was in fact a strongly curved region of space (so motion would involve change in which region was curved). A similar idea was (inconclusively) investigated in general relativity by John Wheeler in the 1970s.

Alternately, if a piece of matter does always take a piece of space with it when it moves, then likely what is going on is that matter is primary, and what is being claimed is that space is just matter. Descartes arrived at this kind of view through the following argument.

First he argued that the only essential property of material things—the property that they had to have in order to be material things at all— was spatial extension. He pointed out that things can be transparent, so color and opacity can't be essential, even if lots of matter is colored and opaque; that bodies can be melted, so matter needn't be hard; that some bodies don't fall down, so weight isn't essential; and so on. According to Descartes, the only property that a body can't fail to have is that of filling some region of space: his idea is that if a body doesn't occupy a finite volume, then it isn't anything (material) at all. (In contemporary physics it's hard to see the point of this discussion. Classifying the types of particle by various properties such as charge and mass is important, but finding a property that all forms of matter must possess doesn't serve any obvious purpose. That said, Descartes is an integral part of the story of how physics came to view space.)

Descartes reached the conclusion that space and matter are one and the same, because he thought that spatial extension (i.e., a volume of a given shape) is also the unique essential property of space: you can't have space without extension and you can't have extension without space. Thus any region has extension and so is a volume of matter (since extension is the essential property of matter) and a volume of space (since extension is the essential property of space), and so the matter of the region *is* the space of the region. And so space is matter, and vice versa. For convenience, let's call it '*space=matter*', pronounced 'spatter'; Descartes's view is that space is matter is space=matter. That is, the chair and the chair-shaped region are not two, coincident things, but one and the same thing—a chair-shaped piece of space=matter. Sometimes we refer to it as a chair, sometimes as the chair's place. (Of course, Descartes owes us an answer to the question 'when is a chair-shaped region of space=matter a chair?', a debt he does not properly discharge.)

So now we ask our question about the motion of matter. Descartes answered 'yes': when I walk across the room, my space=matter goes with me, and some other space=matter (in the form of air) takes my place. Of course this story is quite unexceptional if we emphasize the matter side of the equation, but emphasizing the other side is quite strange: strictly speaking according to Descartes, the piece of space that I occupy goes with me when I walk.

This situation is puzzling, for motion is change of place, which surely means that when I walk I start at one 'piece of space' and end up another. But according to Descartes, the piece of space goes with a body and so does not change: when I walk the Cartesian walk, my space goes with me, so I apparently stay at the same place, walking without moving!

Descartes's answer is that we should not understand the 'same place' to mean the 'same piece of space'. Instead, remaining in the self-same place does not require being composed of the self-same piece of space=matter but instead means maintaining position *relative to some reference bodies*, some things that we choose to take as being at rest. Hence for Descartes, changing place does not mean changing what I'm made of—one piece of space=matter for another—but simply moving relative to other things. So when I walk across the room, I do move in the sense that my space=matter changes position relative to the walls.

Of course, the walls are just one choice among many. I also move relative to the Earth itself, north-south and east-west; and relative to a plane flying overhead, but in a different way, since it is moving relative to the Earth; and in a different way again relative to the fixed stars; and so on. The point is that in principle anything can be taken as a reference body; but since the different possible reference bodies move relative to one another, a given body moves in different ways relative to different choices.

Now it immediately follows from Descartes's identification of space and matter that a vacuum is impossible, for a vacuum is a region of space devoid of matter, but any region of space is extended and so is material! Any attempt to make a vacuum can at best create a region of very low friction, highly mobile matter, whose presence is very hard to detect!

And so the complete picture is of a universe completely full of matter (space=matter) with its parts in relative motion, with no distinct thing— 'space'—at all. A good model for this system is a swirling bowl of water and ice. The water (liquid and frozen) is the analog of space=matter, and all motions have to be understood as the relative motions of parts of water, for instance between two ice cubes. And of course this is exactly the kind of picture we saw in chapter 1 when we discussed Descartes's vortex model of the solar system. Now we know where it came from.

Descartes's view may sound rather fanciful, but is it any stranger than the world of strings that we touched on earlier? Anyway, Descartes developed it not as a flight of fancy, but because he thought this worldview would help him do important work in physics, such as explaining the motions of the planets. In fact Descartes's theory was a complete disaster, not just because it's simply wrong, but also because it is hard to show definitively that it is fatally flawed. Many explanations in terms of the 'fluid' universe sound quite plausible—like the vortex theory of the solar system—and it takes mathematics exceeding that available to Descartes to show that they cannot be right. (Later we will consider some of the more obvious problems.)

9.2 RELATIONAL SPACE

A second response to the space–body puzzle is simply to deny that space is, strictly speaking, a real, physical thing at all (whereas Descartes certainly holds space to be something physical, namely matter); instead it is a logical or mental 'construct' from matter. When we think or talk about or visualize space (and bodies and their positions and motions in space), our ideas, words, and images should not be taken literally, but elliptically, as a convenient shorthand for thinking, talking, and visualizing the complex properties and behaviors of bodies.

The classic exponent of this philosophy was the German mathematician and philosopher Gottfried Leibniz (1646–1716). He laid out his idea in a wonderful correspondence with Samuel Clarke, an important follower of Newton (Newton had some hand in Clarke's side of the correspondence).

Our awareness of space comes from our experiences of things lying at various distances from one another and moving relative to one another (there's an affinity with Poincaré's views). However, we're usually aware of some things that we take to be at rest: for instance, as we watch people walking about or birds flying, we usually take the Earth (or at least the part we see around us) to be at rest. And so we come to think of identifiable 'places'—the southwest corner of Halsted and Harrison Streets, say—where first one thing and then another—me then you, say—can be found. Finally, we think of the collection of all such places, fixed relative to bodies we take to be stationary; together they form our conception of space. That is, according to Leibniz, space is not something in its own right, but something that we 'construct' from the relative positions of material things: we identify fixed places relative to stationary objects and then mentally put them together to make space.

Leibniz, however, claims that we tend to lose track of the fact that places are always places relative to other bodies, so that we mistakenly start to think of the places, and space itself, as things in their own right. As an analogy, he suggests thinking of a family and all the relations that the members have to one another—mothers, aunts, second cousins, and so on. One can describe all these relations by drawing a family tree, but it would be quite wrong to think that this drawing showed anything more than the relations between people, that there is something other than what *the lines of the tree themselves* depict. Similarly, if we drew a map of objects in space, we should not think that we are indicating (or describing) anything more than relative positions: the figures on the paper show the relative positions of bodies, but the paper itself does not depict anything. For obvious reasons, Leibniz's account of space has become known as the 'relational' account of space.

Although Leibniz agrees with Descartes that motion and position are relational, their accounts of space are crucially different. Taking space to be matter, as Descartes did, is not to take it to be a construct, but something in its own right; for Descartes the paper of the map does represent something, namely space=matter. We can also see the difference between Leibniz and Descartes by noting that the vacuum is conceptually possible in Leibniz's relational space: it is conceivable that there are places relative to a reference object that could be occupied but that in fact are not. (Though in fact Leibniz believed God would not allow a vacuum.)

9.3 ABSOLUTE SPACE

Against the Cartesian and relational accounts of space, Newton argued that space is 'absolute' (note that his arguments were against relational space in general, and not Leibniz in particular; Newton made his case thirty years before Leibniz corresponded with Clarke). We can get a grip on Newton's idea, and contrast it to Leibniz's, by considering an analogous case: that of absolute and relative accounts of right and wrong.

Some people hold ethical values to be *relative*, to a particular culture, say. Some (often conveniently for their political goals) say that while it is wrong to deny citizens in the West free speech and political freedom, some developing countries have different ethical standards that don't recognize such human rights at all.

Alternatively, one could instead hold right and wrong to be *absolute*, so that while different societies might have different views of right and wrong, what is actually right and wrong is independent of their views. Perhaps countries that think there are no human rights are just wrong. If so, what grounds right and wrong: what could make an action wrong other than the moral standards of some particular society? One popular (though problematic: see Plato's *Euthyphro*) answer is that it's wrong relative to the moral code laid down by God. For instance, perhaps all people are endowed with unalienable rights by their creator, regardless of who thinks they can be denied. More generally, a typical understanding of absolute right and wrong is that of right and wrong *relative to some very special standard*, one that transcends this or that human standard.

Analogously, Newton argued that in addition to the positions and motions a body might have relative to other bodies, it would also have a unique absolute place and motion: though a person sitting in a flying plane is at rest relative to the plane, is moving 350 mph relative to the Earth, and is moving 700 mph relative to an oncoming plane, according to Newton she would have some absolute speed, likely different from all these. And further, the same standard would apply to all bodies: for instance, the other people sitting on the plane would have the same absolute speed, while the Earth would have an absolute speed that differed from that by 350 mph.

He argued for such absolute places and motions because he took space to be something distinct from matter, something fixed in which all matter was located: he took the absolute place of our passenger to be the very part of space occupied, and her absolute speed to be the rate at which she changes her absolute position. And so, just as absolute right and wrong might be relative to God, Newton reckoned absolute position and motion to be relative to space itself—or to use his term, relative to *absolute space*–transcending positions or motions relative to this or that body. (Earlier it was mentioned that Newton toyed with the idea of matter as impenetrable space; my reading is that the view described now is his 'official' view, though some disagree.)

The disadvantage of 'absolutism' compared to relationism is that it postulates the existence of a rather strange—physical but immaterial, penetrable, inactive—object. But why should we think that such an entity exists? Further, it is an operating principle of science that the simplest theory, postulating the fewest entities, is to be preferred; scientists often call this principle 'Occam's razor', but I prefer the engineers' 'K.I.S.S.', or 'keep it simple, stupid' principle! Newton subscribed to such an injunction, so why did he see the need for absolute space?

One of Newton's motivations for postulating absolute space was that his physics (and indeed that of his contemporaries) could not work if motion were purely relative, but that it could if motion were absolute. More precisely, the problem is that his laws make no sense if 'motion' is understood to mean 'motion relative to any old reference body', because such relative motion is equivocal, depending on the choice of reference body. To demonstrate the point, he considered an example that his contemporaries considered extremely important: an object in rotation.

Imagine a bucket, half full of water, hanging by a rope. If the bucket is spun, friction with the bucket will start the water spinning, and as it spins faster its surface will become concave, and the water will creep up the sides until water and bucket spin at the same speed and the water stays at a fixed height (see figure 9.1). (What does this setup remind you of? The water spinning in Newton's bucket is Descartes's vortex solar system in miniature, so if Descartes cannot account for Newton's experiment, then his account of the motions of the planets fails too.)

Why does all this happen? Recall that to make an object follow an orbital path, a force must be exerted on it toward the center of the orbit: the Earth orbits about the sun because of its gravitational attraction to the sun. In the final state of the bucket, immediately to the outside of any

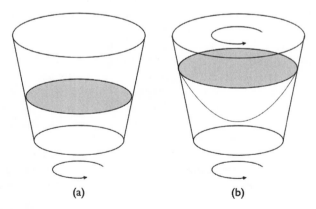

Figure 9.1 (a) A bucket hanging from a rope is filled with water and set spinning. (b) Eventually friction with the sides causes the water to rotate at the same rate, and as it does the surface of the water curves up the sides of the bucket.

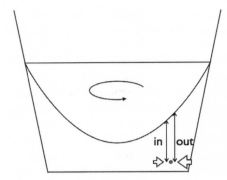

Figure 9.2 When the water is spinning, it is piled up higher to the outside of a water molecule than to the inside. The water column to the outside thus weighs more and exerts greater pressure on the molecule, as indicated by the arrows.

molecule, the water is piled up higher than immediately to the inside (see figure 9.2). The higher column of water weighs more than the lower and thus exerts a greater pressure on the molecule; because of the difference in pressures to the outside and inside, there is a net force on each molecule toward the center. The same effect explains why your tea spills from your cup when you stir it too vigorously: the water can't pile up high enough to keep the tea in the cup!

That is, the faster something rotates, the greater the forces toward the center must be to keep in its orbit. Since the forces in the bucket (and teacup) depend on the slope of the water, the faster the water spins, the higher the water must rise. Newton pointed out that the height to which the water rises thus provides a quantitative measure of how fast the water is spinning. Moreover, since the water can reach only one height, there must be some *unique* rate at which it spins.

But of course the water rotates at different rates relative to different things: in the final stage of the experiment, the water doesn't rotate at all relative to the bucket, at the same rate as the bucket relative to the Earth, faster relative to someone running around the bucket in the opposite direction, even faster relative to the stars of our galaxy (because of the Earth's motion), and so on. If rotation were purely relative, each relative rotation, being at a different rate, would entail a different height for the water; but of course it's impossible for the water to be at many different heights at once. Thus there is, Newton concludes, some absolute rotation that transcends this or that rotation relative to other bodies; he proposes that it is rotation relative to absolute space.

Another analogy might be helpful here (though it takes us somewhat beyond Newton's argument). Imagine a 'Through the Looking Glass' game of chess in which the board is infinite and all the pieces can move at once rather then one at a time. Suppose that toward the end of the game the only pieces left are a knight and a bishop, and they move (simultaneously) from adjacent squares to diagonally touching squares.

How do we tell whether the move was in accordance with the rules or not (whether the bishop moved diagonally and the knight in its L-shaped

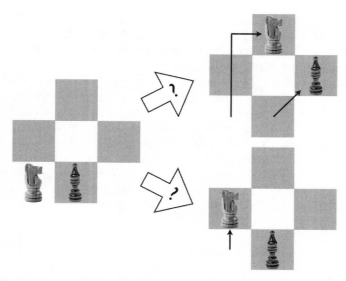

Figure 9.3 Did the bishop and knight move legally (top) or illegally (bottom)? It is impossible to say from their relations to each other, because they are the same either way.

path)? Clearly there is a legal way for the move to take place, but it could also occur in many illegal ways. The knight could simply move one square, or both could move some very large distance to the right, and so on: either way the relative positions of bishop and knight are the same. (See figure 9.3.)

One can tell whether a move is allowed only by taking into account the *board itself*, not simply the relative motions of the pieces. Whether the bishop has moved along a diagonal and the knight in an L depends on their motion relative to the board itself; without the board, the rules make no sense. Similarly, the Newtonian (and Cartesian and Leibnizian) laws of physics, which determine how bodies move, apparently make no sense without *space itself*. Just like the chess board, absolute space defines what it is to move in a straight line or in a circle, and so makes sense of the rules (or rather laws) of motion. In particular, according to Newton the laws determine the height to which the water rises, given the rate of rotation *relative to absolute space*.

Opponents of absolute space often respond that while some rotation of the water has to be privileged as the one correlated with the particular height of the water, it needn't be absolute rotation; it could instead be rotation relative to some privileged *material* object. (Analogously, one could formulate new rules of chess in which allowed moves were all defined relative to the red queen, say.)

In fact, Descartes proposed a view of this kind: he claimed that of all the relative motions of an object, its 'truest' motion was that relative to its immediate surroundings. But Newton's example (intentionally) shows that rotation in this sense is not the same as that in the explanation of the convex surface of the water: at the end the water has climbed up the sides, but it isn't rotating in the bucket at all.

Then again, in the nineteenth century, Ernst Mach proposed that the height to which the water climbs depends on motion relative to the center of mass of the universe. The problem here is to understand how the distribution of distant galaxies, nebulae, and so on leads to a standard of rotation on Earth; if there is some force involved, then what is it? Gravity has no such effect, and Mach never proposed a theory to explain it. Faced with a choice between Newton's physics and absolute space, and a promissory note of a new theory, we should surely accept Newton's position, at least provisionally.

9.4 RELATIONAL SPACE REDUX

There is, however, an alternative, first clearly proposed by James Thomson in 1883, which answers Newton's argument. To see the idea, think about the chess analogy again: we actually could play without a board if the rules stated that a rearrangement of the relative positions of the pieces is allowed as long as it could be produced by legal moves on a board *if there were one*. So when one is playing without a board, in order to tell whether a move is legal, one must imagine the initial arrangement of pieces on a *hypothetical* board and see if the final arrangement could be obtained by legal moves on that board; thus no actual board at all is needed to make sense of the rules.

Analogously, the relationist says that the laws that nature uses to determine which motions are possible (the 'rules of the game') do not require the existence of space itself (the 'board'), but instead require only a hypothetical fixed space to reckon motions. It is perfectly possible to interpret Newton's laws in this way without requiring absolute space. Because of the K.I.S.S. principle, I believe that we should so interpret them. Let's see how.

'Hypothetical fixed space' can be understood in terms of 'frames of reference'. Pick any object as a reference body, and we can give coordinates to any point in space by saying how far in front (or back), to the left (or right), and above (or below) it is. Such a system of coordinates is a 'reference frame', relative to the reference body; it's a system of labels or names for different places in space.

We can make the idea more general by saying that the reference object can be offset in the frame; so the coordinates of a point might be given by the distance in front +10 m, to the left −20 km, and above +1 lightyear. And we can make the coordinates even more general by saying that the reference object could be moving in the frame. Perhaps that first coordinate is now given by the distance from the reference object

+10 m/s × the time after 12 P.M., so that the reference object is moving away from the origin at 10 m/s. Such coordinates are just alternative ways of labeling places relative to the reference body.

The key point is that the laws of physics do not work in just any reference frame. For instance, they don't work in a frame in which Newton's bucket is always at rest, for in such a frame the water it contains is initially moving and so should have a curved surface according to the laws—which of course it does not. (And vice versa at the final stage of the experiment.) However, relative to any reference object we can define some, probably moving, reference frame in which they do hold. Such a frame is called an 'inertial frame': for instance, a frame in which the bucket rotates. To put the whole thing in a nutshell, the kind of motion governed by the laws of physics is motion in an inertial relative reference frame.

Obviously there is a significant difference between inertial and noninertial frames; some ways of labeling places are singled out. The question of the nature of space turns on understanding this difference.

According to Newton, the kind of motion talked about by his laws is motion with respect to absolute space; in his view, then, coordinates fixed in absolute space form an inertial frame. (The converse is not quite true; Newton showed that any frame moving at a constant speed in a straight line without rotating would also be inertial.) For him, then, the inertial frames are singled out *because* they label absolute space properly.

But why say that? Instead say that it is simply the laws themselves that single out the inertial frames, that the inertial frames are singled out because they label points in a way that makes the laws work. Coordinates are just ways of naming places, and we have no need to add absolute space to the story in order understand why some ways of naming are singled out. All we need are the frames relative to reference bodies, and the fact that the laws work in some and not in others.

In this regard, dispensing with absolute space is quite different from, say, trying to dispense with atoms. That is, one might gloss what has just been said as 'there is no absolute space, but motion is motion *as if* in absolute space'. Then one might compare that view to saying that 'there are no atoms, but things behave as if there were'—that atoms are just fictional. But this latter view looks suspicious, because we just can't do science without talking about atoms, suggesting they are real, even if we want to pretend they are not. But doesn't what goes for atoms go for absolute space too?

Our discussion shows it does not, for absolute space does not play a role like atoms in our theories, explaining why this or that happens. Instead it merely provides an understanding of what is special about certain ways of naming locations, but it's an understanding we don't need. We can say instead that inertial frames are nothing more or less than those frames in which the laws are true. If you like, the attitude is that we do not need to appeal to absolute space to explain why Newton's laws hold in some frames; instead that is just a basic fact about the world, not the kind of thing that needs explaining.

It seems to me that after all, nothing more need be said about inertial frames or motion, and adding absolute space to the story merely adds an unnecessary complication. That said, I'm sure this discussion will not settle the issue and that further considerations can be brought to bear. We will rest content here having seen how the relationist can respond to Newton.

9.5 WHAT PHYSICS AND PHILOSOPHY CAN TEACH EACH OTHER

This chapter illustrates nicely how the interaction between physics and philosophy can be a two-way street. We started with a philosophical question about the nature of space and a (questionable) philosophical argument that an object and the space it occupies are two different things—an argument from language. In response we considered Leibniz's account of the language of space, in which space is nothing but a construct.

But we also saw how the issue came up in attempts to construct a theory of motion. In particular, Newton argued that relational accounts like those of Leibniz (and Descartes) could not make sense of 'motion' in the sense discussed by the laws. He concluded that space was indeed something distinct from matter. Newton's achievement here is twofold. First, he gave a clear and adequate account of what it means in physics for something to move. Second, he brought considerations from physics to bear on the philosophical question concerning the nature of space.

In contrast, we saw, with Thomson, that we can replace absolute space with a frame in which the laws are correct: an inertial frame. He offers, then, an alternative account of motion: motion with respect, not to space itself, but to inertial frames. This is the kind of analysis of what a concept means and how it is used within a theory that we have seen before, from Zeno for instance. As was said earlier, such analysis is exactly what philosophy contributes to physics.

I have stressed how Newton argued that space is an absolute something, other than matter, in which case Thomson's account supports an opposing, relational account. However, that picture is a bit unfair to the view that Newton laid out in the book that he intended for physicists, his *Mathematical Principles of Natural Philosophy*, in which it is discovered that gravity holds the planets in their orbits. In that work, all absolute space does is provide a frame for the laws, and its distinctness from matter is not stressed. In other words, it is not much more than an inertial frame. Moreover, when Newton actually calculates the motions of the planets, he looks not directly look for absolute space but rather for a frame in which the laws hold; he takes it to be (to a high approximation) the frame in which the stars are at rest.

The point is that Newton understood what role absolute space was to play in physics, and it was much the same as that proposed by Thomson for inertial frames. However, Newton also clearly did think of absolute space as more than just an inertial frame, and he elaborated elsewhere just what that was. So Newton engaged with the philosophy of space and motion in two stages: he understood what motion amounted to in physics, but he also thought that story supported a particular philosophical view about the nature of space. In the former stage we see philosophy illuminating physics, while in the latter, an attempt to draw philosophical conclusions from physics.

We've reached the end of our discussion of space; in the following chapter we turn to time. So let's take a moment to look back at the landscape we have traversed. We've seen how various formal disciplines have developed that capture relevant concepts in physics: calculus describes variation, and topology and geometry make precise the idea of the shape of space. I've sketched an introduction to these disciplines in order to make clear the shape of some of the most fundamental mathematical ideas. These ideas will continue to serve us as we progress.

As we've gone on we've increasingly wondered about space at its most basic level: especially, in the last two chapters, what are we really talking about when we talk about the geometry of space, and how does space relate or compare to material things? Here I have defended the idea that we should not take talk about space too literally, that what space really boils down to is the spatial properties of material things. I have thus taken, and argued for, a controversial philosophical stance about the meaning of physics. This view will color some more issues as we progress (chapter 16, for instance, which discusses the nature of handedness), though we will also move onto further topics as we seek to understand how physics and philosophy talk.

Further Readings

Descartes's view of space and matter are best discussed in the *Principles of Philosophy*, predominantly in Book II. Relevant extracts—and my commentary, which of course I strongly recommend—can be found in my *Space from Zeno to Einstein: Classic Readings with a Contemporary Commentary* (MIT Press, 1999).

The Leibniz-Clarke Correspondence is published in full by Manchester University Press (1998), but the parts most relevant to space are also reproduced in my collection.

So is the 'Scholium to the Definitions', from Newton's *Mathematical Principles of Natural Philosophy* (the *'Principia'*), and his unpublished essay, *De Gravitatione*, in which he describes his philosophy of absolute space.

The great exponent of the logical construct was Bertrand Russell, who applied the method to a variety of philosophical problems. The best starting point in his philosophy is the terrific *Problems of Philosophy* (Oxford University Press, 1998; originally published in 1912 but still a great introduction to philosophy).

Thomson's essay appeared in the *Proceedings of the Royal Society of Edinburgh*. His work is discussed in Julian Barbour's *The End of Time: The Next Revolution in Physics* (Oxford University Press, 1999), which pursues a different relationist account of space.

William Kingdon Clifford's ideas appeared in his posthumous book *Common Sense of the Exact Sciences* (Thoemmes Press, 2001), which is a wonderful popularization of science from the turn of the previous century.

10

Time

This book is organized into broad (overlapping) themes; we started with motion and then moved onto space. In this chapter we turn to time.

What is time? If we try to picture it, the natural thing to do is to draw a line, just as we do for a spatial dimension; in fact, that is exactly what we do when we plot a graph of something changing over time. But problems arise if we think of time as something like a spatial dimension in this way, because our experiences and common sense attribute radically different features to space and time. An incomplete survey of these features includes: time passes, we have no control over our position in time, time is directed from past to future, we have memories only of the past, we can affect only the future, time can have a beginning and end (though surely there's a temporal version of Archytas's argument), and so on.

Our goal is to see how physics pictures time—rather like space, but not exactly—and how that image can be reconciled with some of most striking of these features.

10.1 TIME VERSUS SPACE

Imagine the whole universe right now, at this very moment (you can observe a bit of it around you, but picture it stretching out as far as it goes, perhaps to infinity); you're imagining everything that there is, 'frozen' in a purely spatial instant in the history of the universe. I find it easier to leave out a spatial dimension in my imagination, and picture the universe distributed over a huge two-dimensional sheet. (You may know that in relativity theory there is no absolute present like this. Let's ignore that complication for now; we will return to it in chapter 14.)

Similarly, one can imagine other instants: one second after the big bang, the moment John F. Kennedy was shot, 7 A.M. this morning, exactly 100 million years in the future, and so on.

And now imagine the entire collection of these instants all together, arranged in temporal order: leaving out one spatial dimension for ease of visualization, picture a stack of two-dimensional arrangements of everything that there is. What you are picturing is the entirety of the universe

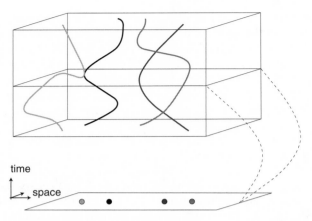

Figure 10.1 A stack of three-dimensional (or here two-dimensional for illustrative purposes) spatial slices, each containing the arrangement in space of everything that exists at that time. I've pulled one out. The lines represent the 'worldlines' of bodies; the dots represent them at an instant.

(less one spatial dimension) in all of space and time; it's illustrated (schematically!) in figure 10.1.

Since time is one of the dimensions, objects are shown extended not only in space, but also in time. Suppose, for instance, that the line starting rightmost represents a ticking clock. Each of the slices contains a clock-at-an-instant, occupying a region of space, reading a particular time (since the clock is pictured occupying time as well as space, we can consider temporal as well as spatial slices). The series of these clock-at-an-instants constitutes the whole history of the clock, all the different times it ever reads, and such a path through space and time we call a 'worldline'.

In this image (although I've drawn axes to indicate how the stacking is done), time is just another dimension, at right angles to the spatial dimensions but not obviously distinguished from space in any interesting way. The question that faces us is whether the image is complete or whether time and space are distinguished somehow.

The first thing to notice is that they are not treated on a par by physics. For one thing, the laws of physics effectively say, 'Tell me how things are arranged in space now, and I will tell you how they will be arranged at other times.' That is, pick a point in one dimension—that of time—describe the state of the world in the remaining three dimensions at that point, and the laws determine (at least probabilistically) the state at other points of time.

What if we pick a point along a spatial dimension and describe the whole world at that point in the remaining three dimensions—two of space and one of time? We would have described the state of a two-dimensional plane of space for its history: past, present and future;

imagine the entire history of the infinite plane that bisects your body back–front. That description is generally *insufficient* to determine everything that happens elsewhere, in front and behind you, to the ends of the universe.

That is, feed the state at a point along the temporal dimension into the laws and they will give you the state at other such points (at least probabilistically); feed in the state at a point along one of the spatial dimensions, and they *won't*. And so the difference marked by the space and time labels is a real physical one. But what is its nature?

One possibility is that the only thing singling out one of the four dimensions of the universe is the role it plays in the laws: the laws say that some dimension will play the role described, and time is just whichever dimension does. If one also thinks of laws as simply descriptions of what happens—as I suggested—then we get the following picture: at base, the universe is composed of things arranged in four dimensions, with no distinctions among them; but the best description of the way they are distributed turns out to single out one dimension, which we call time. That is, at root, the distinction between time and space arises just from the way things are distributed across four interchangeable dimensions.

Alternatively, is the dimension of time distinguished from space by a special property, a basic 'temporality'? Such an idea seems plausible if one thinks that the fact that the laws single out a special dimension stands explaining: because one dimension is special, the laws treat it differently, or that is why they can. I don't intend to debate these competing views, just to present them as important issues. However, my sympathies are with the view that the distinction lies in the laws, because I am skeptical that explanations are needed here; the laws just are that way.

There are other features of time that distinguish it from space (perhaps one wants to suggest that they can explain the role of time in the laws). Especially, one of the known fundamental laws—governing subatomic particles—distinguishes a *direction* to time.

That's unusual, because most of the fundamental laws work the same in both temporal directions. For instance, if they say that two particles will fly apart at some speeds on collision, then they also say that the reverse is possible; if they flew back with their motions reversed and collided, then they would fly apart with their initial motions reversed. (Everyday experience isn't quite like that, because colliding objects lose energy in collisions and heat up. What was said holds of processes involving the most basic physical things.) But with the 'weak' nuclear force, it is not true at all that processes can run in either direction in time: some only ever happen one way.

The first thing to note is that something similar applies to space. An 'irreversible' process is like an 'arrow' pointing from past to future; the impossible reverse would be an arrow pointing in the other direction, from future to past. These arrows are reflections of each other in the time dimension, and the irreversibility of the weak interaction can be thought of as singling out one of these mirror images.

The spatial analog then is if the laws single out one-handedness rather than its mirror image. For example, imagine a law that says marble hands are spontaneously created at a rate of one every hour, and that they are all left: the creation of a right hand is an impossible process. But the weak interaction is exactly like that. It doesn't involve marble hands, but it does allow certain processes of one-handedness and not the other. (The processes involve the interactions of several particles: it's not hard to picture how their trajectories in space could form a figure that is not 'mirror symmetric'. We will return to handedness and mirrors in chapter 16; there you will also find further readings on the weak force.)

Thus with respect to both space and time the weak interaction is 'irreversible' under reflections. Does that mean that, after all, irreversibility is irrelevant to the distinction between space and time? Of course not. We still have the fact that time has a *one-dimensional* arrow while space has a *three-dimensional* handedness. Thus reflection irreversibility separates the four dimensions into a group of three (space) and a group of just one (time); a 'handedness' for each group is picked out.

Neither of these handednesses is represented in our image of space and time either, so they should be included. Time comes with an arrow, and space with a particular handedness to show which processes are possible; let the arrow run from past to future and let the handedness be left. Again we can ask about their significance.

Again, one can argue that they have their full significance just because of the *law* of the weak force; it says all similar processes will have the same handedness, and we are indicating that feature. One could also argue that in some sense time gets its arrow without the law, and that time's arrow helps *explain* the law's asymmetry. I'm going to set this debate aside too; it's important, but I want to address a different issue. (I'm also setting aside for now other reasons one might attribute an arrow to time, and how they might enter into the argument.)

10.2 NOWISM

The usual name for the four-dimensional conception of space and time we have introduced is the 'block universe'; it encompasses all of space, all of time, and absolutely everything that happens, 'everywhere and everywhen'. There is a view that this picture drastically misrepresents time by leaving out something crucial: which moment is the *present* one, which one is 'now'.

What is proposed is a further difference between time and space. 'Now' distinguishes a moment from other times (the past and future) and 'here' distinguishes a location from other places, but we tend to think of them differently. 'Here' is just where you are, while 'now' seems to have more objective significance than the time relative to *you*. If that's right, then while the block represents everything about space (without marking out any place as 'here'), it gives an incomplete picture of time because it fails to represent the fact that some moment is the present. Because it sees the present as a crucial element to time, with no spatial analog, I'll call this view 'nowism'.

We'll spell out some more concrete ideas under this heading. We'll see that the nowist's present cannot be added to the block, so I will call the other side of this debate 'the block view of time'. 'Blockism' makes two claims. First, the block accurately represents the view of time held by contemporary physics. We have seen how time is singled out and given an arrow in the block; other structures could be added if physics demanded. Further, I will just state that physics requires no privileging of one moment as *the* present. It is the second blockist claim that will occupy us mostly: there are no nonphysical grounds for a nonblock present. According to the block view, the image of time according to physics, and captured in the block, is entirely satisfactory.

Sometimes 'blockism' includes the stronger idea that space and time are at root equivalent, but I don't intend that here. We have differentiated space and time in the block and treated the nature of that difference as a further question. Similarly, 'nowism' is a new term, because existing ones are used in too many ways. ('Isms' should be handled with care: one label comes to cover a variety of different ideas, and it becomes easy to slip between them without realizing that different issues are at stake. We'll be careful to stick to well-delineated issues.)

Why would anyone be dissatisfied with the block? First, perhaps the present is required to make sense of change; after all, we say something changes when it is one way at one time—in one present—and another at a second, in a later present. Imagine watching a clock: now it reads exactly 12, now a second after. That analysis seems to involve reference to which moment is the present, and indeed a change in which moment is the present, so don't we need to add that information to the block?

There are a couple of things to say. The first is that the idea of the present makes sense in the block, as it better had (just, according to the nowist, not the right sense). When someone says 'now is the winter of our discontent made glorious summer' or 'Putin is the present prime minister of Russia', perhaps all they mean is that these things are the case *at the time they are said*. If so, 'now' is being used just like 'here', to refer to one's location in the block—in the former case one's temporal location, and in the latter one's spatial location. And just as this use of 'here' makes sense in the block, so does this use of 'now'. But the nowist wants more; 'here' may just be a matter of where you are, but 'now' isn't just relative. That's a puzzling idea, but then again 'now' does feel objective in a way that 'here' does not, so what is going on?

The second point is that our analysis of change does not, after all, require any nonrelative sense of the present. We imagined that it could be truly said of different times at two different moments that 'the clock reads so-and-so now'. All that is required is that the clock read the appropriate times at the moments the sentences are said. But reading the appropriate times at the appropriate moments is just the 'at–at' theory of motion of chapter 3 applied to the clock. Motion is just a special kind of change, so in general we can say that in the block any change is a matter of being in different states at different times.

Those who hold the block view are typically satisfied with that response (though more resources are available if necessary; perhaps time needs its arrow to fully explicate change). So consider another issue. Much of the intuitive appeal of nowism is the difference between our perceptions of time and space, the feeling that time is the only dimension that *passes*. This feeling has many facets: that the past is out of reach (harsh words can never be unspoken), or the uncontrollable rush toward some dreaded event (that embarrassing meeting), but the aspect on which I want to concentrate is the *dynamism* of our experiences.

I have in mind the powerful sensation of time flowing, of the immediate future becoming the present and the present becoming the immediate past, of a changing world in constant flux. In contrast, think of the way things appear to vary through space: there's furniture around here, cars a bit further away, then buildings, and sky above it all. But while I can see how things are different in different spatial regions, I have (outside of vertigo) no analogous sense of far *becoming* near: I experience *change* over time but only difference and variation in space.

This difference motivates nowism because it is tempting to think of the experience as just being that of a present—in a nowist, nonrelative sense—somehow in motion: 'We know that there is a nonrelative now because we constantly experience its motion.' As I intend the designation, the nowist holds not only that there is a present in a sense not captured in the block, but that it moves. (We'll see very soon how problematic this idea of 'motion' is.) The blockist rejects that claim but is left to give some account of the experienced difference between space and time. That job will be undertaken in the final section of this chapter. First I'll make good on the claim that nowism involves more than a modification of the block, that it requires a wholly different conception of time. Doing so will also help us clarify how the nowist conceives of the present, and what the difficulties are for nowism.

10.3 A MOVING NOW?

Suppose that we try to explicate nowism as adding something—some kind of additional structure—to the block. What sort of structure? Adding more material (or mental) things won't make time move and neither will more space or more instants; those kinds of things are already in the block. What then if one instant in the block is picked out as being

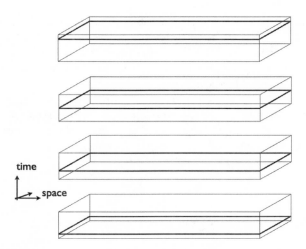

Figure 10.2 A block of spatial slices, only one of which is illuminated at any time.

the present? We can visualize this proposal by picturing a stack of two-dimensional spatial arrangements (like the frames of a movie stacked up) as a model of the actual (four-dimensional) block universe. Then picture one of these 'slices' as singled out by being illuminated (representing the present).

Picking out a single instant of the block as the present doesn't make it move, and doesn't satisfy the intuition that we directly experience a dynamic passage of time. So let's try to cash out the nowist view by imagining the slices in the stack model of the block illuminating then dimming in sequence. Analogously, each instant in the block is the present—is 'illuminated'—at some point (see figure 10.2).

The image certainly makes sense, but it is much harder to make sense of a present moving through a real block universe. The problem is that the change in which slice is illuminated is a change over time: at one instant some slice is illuminated, and at a later instant it's another. Following the analogy through, the change in which instant is the present will also be a change over time: at one instant some instant is the present, and at a later instant it's another.

That sounds fine, a rather trivial truism even, but the problem for the nowist is that it is also true in the plain block universe! For instance, at noon last Tuesday one instant was the present, but another is the present now! The truism is, unsurprisingly, true of the block! So while a stack of unilluminated slices is different from a stack of slices illuminating in sequence, a universe of all instants of time is a universe in which every instant gets to be the present at some time, namely at itself! And so this analogy fails to clarify at all what kind of addition to the block universe a moving present could be.

It's pretty clear what has gone wrong. In the supposed analogy something changes with respect to time, but in the block universe the something that is supposed to change just is time. It is hard to make out how time is supposed to change with respect to itself, in any sense other than it already does in the block universe.

About 100 years ago the philosopher John McTaggart argued that you can't have a moving present in the block. (He further concluded that time must be an illusion; we won't!) His argument contains a number of obscurities, but we will consider just its core, which does (after some holes are filled) do the job.

10.4 McTAGGART'S ARGUMENT

Think again of the stack of spatial slices illuminating in sequence, and suppose for the sake of argument that this model does, after all, somehow represent an intelligible moving present. The root of McTaggart's problem is that the block, by definition, contains *everything* that ever happens, including each instant's being the present in the nowist's sense. So pick any instant in the block; if it isn't the present, then the block doesn't contain everything that happens, because one of the things that happens is that instant's being the present. But the block, by definition, does contain everything that happens, so the instant must be the present. And the same goes for any instant, so they all must be the present. In that case the present does not move at all—in fact it is at all times, leading to the apparently contradictory situation of all times being *simultaneously* the present.

It helps to think through the argument using our analogy. The stack of slices is a model of everything that ever happens, and it models slices being the present by illuminating them. Pick a slice; if it isn't illuminated, then the stack doesn't represent its being present. But then the stack doesn't represent everything that ever happens; and so the slice must be illuminated. And the same argument goes for any slice, so they must all be illuminated, and the illumination doesn't move after all, and indeed the stack represents every instant as being the present, all at once.

McTaggart's argument assumes, perfectly naturally, that the property of 'being the present' is the *same* property for every instant (as each slice is illuminated in the same way). Then, since the block contains everything, all instants end up having the same property; they all end up, contradictorily, being the present. But what if instead there are *different* properties of presentness, one for each instant? In analogy, what if the slices in the stack are illuminated in *different* colors?

Suppose there were another sense of 'being the present': not simply 'being the present' but 'being the present at . . .'. That situation doesn't lead to a contradiction; that every instant is the present at some time does not mean that there is any one time at which they all have that property. Whether we would have a moving present is another question.

The question is how to fill in the ellipses in 'being the present at...'; what is supposed to correspond to the differences in color? Times, it seems: 'being the present at 11:59 P.M. on December 31, 2000' and 'being the present at 12:00 A.M. on January 1, 2001', and so on. The problem with this idea is that instants already have such properties in the block; 11:59 P.M. 12/31/00 is the present at 11:59 P.M. on December 31, 2000 in the block, and 12:01 A.M. 1/1/01 is the present at 12:00 A.M. on January 1, 2001 in the block, and so on. That is, every instant in the block is the present *relative* to the time that it occurs, so filling the ellipses with block times fails to add anything to the block.

What if there were some 'second dimension of time'—we can label it '*TIME*'—distinct from that in the block? Then 11:59 P.M. 12/31/00 is the present when the *TIME* is 11:59 P.M. on December 31, 2000, and 12:01 A.M. 1/1/01 is the present when the *TIME* is 12:00 A.M. on January 1, 2001, and so on. The present 'moves' because different block instants are the present at different *TIME*s.

If you find this idea confusing, I'm sympathetic. But it is clear enough to see what problems it would face if it were made clear. First, physics recognizes no such second time dimension, so the proposal has no scientific support. It also follows that adding a *TIME* dimension gives a new, nonblockist account of time; by 'the block' we mean the image of time in physics, which does not postulate *TIME*. Hence my claim that the moving present cannot be incorporated in the block.

But this alternative picture of time won't fly anyway. First, the idea of dual times is violently counterintuitive: we experience just one time and should not sacrifice that experience in order to save another (dynamism) without good reason. Second, in what sense is *TIME* not just familiar time? The two times are supposed to be in one-to-one correspondence, so it is natural to think that each instant of *TIME* is just the corresponding instant of time. Finally, one cannot answer this question by saying that *TIME* passes of it own accord while time passes only with respect to *TIME*. Saying how *TIME* passes is just the same kind of problem as saying how time passes.

Thus the nowist, in response to McTaggart's problem, needs a new picture of time, but not this one. What many nowists propose is a view according to which the block does not exist; what exists is restricted to the present alone. That is, we should picture time not as being like a stack of slices illuminating in series, but somehow as a series of individual slices.

One problem with this suggestion is understanding how the slices form a series; it's natural to say 'in time', but that suggestion seems to make the series the block again, with the problems we've already seen. To avoid repeating ourselves, we'll put those worries aside and assume (generously) that the proposal makes sense. Then, a second problem is that this account fails to explain our experience of dynamic change any better than does the block universe.

At any time, the unique instant that exists according to the nowist is just the unique instant of the complete block which is the present according to the block theory. But if two things are just alike at every instant of time, then they must be changing in just the same way. And so in whatever sense 'unique existence' moves according to the nowist, 'presentness' must 'move' through the block. But according to the nowist, the latter is inadequate for our sense of time passing, making it very hard to see how the former, identical kind of motion could either. If there's no moving present in the block, then there's no moving present in a sequence of uniquely existing instants either.

Maybe that reasoning is a little hasty. The nowist might reply that our experience of different instants *coming into existence* explains our dynamical sense of the passage of time in a way that mere *being present* cannot. That is, our sense of dynamism requires the unique *existence* of the present: its presentness does not suffice. But if that is what is proposed, then the nowist is far from our normal understanding of our experiences. They are normally explained in terms of the effect of the world on us, and how we react to those influences. There's no part of those explanations that requires also mentioning that the things causing our experiences exist.

When the nowists appeal to dynamism, it seems that they don't really mean that a moving present is needed to *explain* it, in the normal way we explain experiences. Instead their argument is a more direct one, that we somehow sense the present moving in a way that requires no explanatory story: we experience it 'immediately', we might say.

10.5 PASSING TIME IN A BLOCK UNIVERSE

Consider what the great physicist-philosopher Arthur Eddington had to say about the 'dynamic' quality of our experiences of time:

> It is clearly not sufficient that [a process in the world] should deliver an impulse at the end of a nerve, leaving the mind to create in response to this stimulus the fancy that it is turning the wheel of a cinematograph . . . we must regard the feeling of "becoming" as . . . a true mental insight into the physical condition that determines it . . . because in this case the elaborate nerve mechanism does not intervene.

Eddington did not advocate nowism in our sense. He (unhelpfully) analogizes what I called the experience of dynamism to the feeling of cranking a hand-driven movie projector, but he doesn't think it's an experience of a 'moving' present. Rather it is of a 'one-way texture' to time, like a cat's fur, running from past to future, another arrow of time.

However, he shares with many nowists the view that the experience of dynamism involves not just experience of the things that happen in time, *but of time itself*—for the nowist, of the motion of the present. Here I want to explain, on the contrary, how the experience of dynamism can be understood as being of events in the ordinary way, not of some feature of time itself. Doing so will undermine an important intuitive motivation for the moving present.

As I said, a large part of the experience of dynamism lies in the differences in the experiences of change over time *versus* variation across space. Of the various changes we experience, change of place—motion— is probably the most ubiquitous (at least of those lying outside ourselves). I'm going to explain why our experience of motion is different from any experience of analogous variations in space. That explanation will not prove that no aspect of experience is of time itself, but it will show that a significant aspect need not be understood that way, and so suggest that all aspects can be addressed similarly.

Suppose that you want to detect motion in the visual field of an artificial eye: something that focuses an image on a grid of light-sensitive cells (they signal 'yes' when illuminated and 'no' when not). You could take pairs of cells, feed the signal of the first into a short delay loop and then, with the output of the second, into a device that will signal 'yes' only if two yeses are input to it. This simple system will signal 'yes' when light falls on the first cell *then* the second, which will happen if a bright object moves across the field of vision. It detects motion in a line from the first cell to the second (so in any direction, for the right choice of cells).

The human brain contains something along the lines of this model detector, at the back of the head, in the 'MT' region of the visual cortex, taking inputs from the retina. The mechanism is considerably more sophisticated but close enough for our purposes (one important difference is that it detects the motion of light-dark edges, not bright regions). The effect of these detectors can be felt in some striking illusions. Let me explain.

Suppose there are two of the model detectors: one detecting leftward motion and one rightward. If a light spot moves back and forth sideways across the artificial eye, the detectors signal yes alternately, and back-and-forth motion is detected. Now suppose that after the right cell is illuminated there is a brief pause during which there is no light, and then the light moves left to right again. Although right-then-left illumination occurs, the pause means that the time delay is too long for our detector, and no leftward motion is detected: our system (mistakenly) detects something constantly moving to the right.

Although the detectors in our brains are more sophisticated, they can be manipulated in a similar way. If you watch a loop of two movie frames of something moving to the right, both leftward and rightward detectors will fire; you will perceive back-and-forth motion. If you watch a loop of the same two frames plus the negative of the first, then only the rightward detectors fire, because the negative foils the detection of light-dark edges leftward—*so you perceive something moving rightward while not actually getting anywhere*! (It's at the same place at the start of each loop.)

It's a strange experience, but you really need to see it: the *Further Readings* shows how. (In a similar audio effect, a repeated tone seems to be always rising: search the web for 'Shepard Tone'.)

The illusions isolate that feature of experience corresponding to the brain's motion detection mechanism, by making it fire independently of the usual concomitant change of place. Thus we learn that our usual, non-illusory experiences have two aspects: of objects somewhere *and* moving a certain way. (It's as if at each moment moving objects come flagged with an arrow showing how they are moving.) It's far from intuitive that things are this way. It's easy to suppose instead that to experience motion is to have series of experiences of something at different places. But what we have learned is that even a single moment in the series involves the experience of motion.

I claim that the feature isolated contains all the 'changeyness' of the experience of motion, which leads people to propose that we have direct perception of the passage of time. The claim is in part based on introspection of the illusory perceptions (so *you* have to experience them). But more important, we identified the character of the experience of change *in contrast* to the experience of spatial variation. The human motion-detection mechanism explains that contrast, because it detects motion, not spatial variation, producing experiences of motion, not variation.

The point is *not* that just because our experiences of motion can be wrong, they aren't reliable indicators of a moving present. The point is that we know that mechanisms in the brain detect motion, and the illusions show how those mechanisms correlate with our characteristic experience of the 'changeyness' of motion. This striking aspect of experience, which many have thought could be only of some feature of time itself, can be understood without postulating anything beyond the mechanisms. Since this account is of a piece with the rest of our understanding of the mind, it seems far preferable.

Sometimes nowists say that if the block view of time is correct, then our experience of the passage of time must be an illusion. If I'm right, then they are wrong, at least for the aspect of that experience under discussion. If something moves across my field of view, triggering my motion detectors so that I perceive motion, then that's no illusion; the thing really is moving. To think that the experience is of some feature of time itself is not to experience an illusion, but to *misinterpret* that experience. It's akin to seeing someone from behind and mistakenly thinking I know them; there's nothing illusory about their appearance, but I misinterpret its significance.

I have discussed only one aspect of temporal experience, but I have shown how a striking feature can, despite what Eddington says, be understood in terms of the 'delivery of impulses to nerve ends', for that is all the mechanism requires. A convincing response from Eddington or the nowist requires identifying some other aspects of experience and an argument that it cannot be understood in similar terms. Otherwise, our success here is good reason to think that all the characteristic aspects of temporal experience can be similarly understood.

Time is part of the subject matter of physics as well as philosophy, so it's not surprising that its study falls with the philosophy of physics. Our discussion started with the block universe, which is a philosophical image of time, drawn from math and physics: we saw the connection with the at–at theory of change, for instance. But now we have also started discussing perception, for we have come up against the idea that direct experience can also teach us about time, an idea I have resisted.

Further Readings

Three books on the philosophy of time in general which I recommend are Craig Callender (whose account of the space–time distinction I borrowed) and Ralph Edney's *Introducing Time* (Icon Books and Totem Books, 2001), *Travels in Four Dimensions* by Robin Le Poidevin (Oxford University Press, 2003), and Michael Lockwood's *The Labyrinth of Time* (Oxford University Press, 2007). I also love Alan Lightman's little book *Einstein's Dreams* (Warner Books, 1994) for a literary exploration of time. Once again, Brian Greene's *The Fabric of the Cosmos* (Knopf, 2004) is good for a physicist's thoughts on the the block universe (and many other topics).

The quotation from Eddington comes from his *The Nature of the Physical World* (MacMillan, 1928: 89). McTaggart's argument is found in 'The Unreality of Time', first published in 1908 in *Mind*, but widely republished. In his terms, the block is a 'B-series', while the block plus moving present is an 'A-series'.

George Mather is a leading figure in the science of motion perception, and the illusions on his web page will amaze and educate you: http://www.lifesci.sussex.ac.uk/home/George_Mather/Motion/index.html

11

Time and Tralfamadore

In his book *Slaughterhouse Five*, Kurt Vonnegut imagines beings from the planet Tralfamadore who have individual experiences, not of things at a time, but of things extended in time—individual experiences of the histories of things. According to Billy Pilgrim, the novel's protagonist:

> the Universe does not look like a lot of bright little dots to the creatures from Tralfamadore. The creatures can see where each star has been and where it is going, so that the heavens are filled with rarefied, luminous spaghetti. And Tralfamadorians don't see humans as two-legged creatures, either. They see them as great millipedes—"with babies' legs at one end and old people's legs at the other," says Billy Pilgrim.

It's a gripping image of spacetime—so different from the way we experience it—the 'spaghetti' is seen in figure 10.1 and the 'millipedes' are people's worldlines. Our experiences are of things extended in space, while theirs are of things extended in time (and space). (Of course, a *series* of experiences amounts to experience of things extended in time; the point made here is that we can individuate our experiences of things in space at a time, in a way we cannot do the converse; we just don't experience things the way the Tralfamadorians do.) Vonnegut draws attention to an asymmetry in our experience of time and space that is worth investigating; why would our experiences have this character? What does it show about space and time in the block?

A complete explanation of our experience of time would require more than philosophy and some simple physical ideas: theoretical and experimental work in cognitive and neuroscience and most of all a convincing scientific understanding of consciousness (see the suggested readings for some relevant discussion). All that we can attempt is a sketch of an answer, to make it plausible that a block universe does contain the resources we need to understand our experience of time—that the kinds of things that are relevant to what we experience have the kinds of features that can give rise to the singular character of our experience of time. The goal is to investigate at a more basic level what is necessary for the more detailed kind of story discussed at the end of the previous chapter.

11.1 THE MIND'S WORLDLINE

To understand this asymmetry in our experience, we will start by considering the place of our minds in the block. It's reasonable to think that what you experience is determined by what happens in your brain. But for our purposes a weaker claim suffices: that the state of the brain, perhaps in addition to other factors, makes a difference to what is experienced. And of this assumption there can be no doubt; much is known about how the state of the brain corresponds to different aspects of experience. Hence we will take 'the place of your mind' to mean the worldline of your brain.

In the natural ways of comparing lengths and times, people—and hence their brains—occupy regions of the block that are extended much further in time than in space. Light travels at 3×10^8 m/s and sound at about 1 km every 3 s, so according to those measures 1 s compares to either 3×10^8 m or 300 m. Life expectancy in the United States is 78 years, or 2.5×10^9 s: equivalent to about 10^{18} m or 10^{12} m, by the light and sound measures respectively. By either measure our temporal extent is much greater than our spatial extent of a meter or two. Your worldline, and so that of your brain, is a long, narrow tube (with millipede legs) running in the direction of time.

The relevant measure for this discussion, which concerns mental processes, is the speed at which effects propagate through the brain. The slowest signals travel along neurons at 1 m/s; even by that measure, a human life equates to over 10^9 m. (Of course there are slower processes. Before global warming, 'glacial' meant maybe 100 m/year—less for Alpine glaciers—while tectonic plates move at a few centimeters a year, making the average human life nearly 8 km and a couple of meters, respectively. By the latter measure, our worldlines look squat, but these processes are not relevant to the story that we are developing.)

Next, the brain contains a 'representation' of the world around it. Things in the block universe transmit 'signals', especially reflected light or sound, that affect our sense organs, especially our eyes and ears. In turn, they produce further signals that serve as 'inputs' to the brain, which uses them to construct the representation—a kind of 'picture' of the things that produced the signals.

Of course I don't mean a literal picture. A more reasonable metaphor is the way that light signals from objects affect the light detector (CCD) of a digital camera, which provides inputs that change the state of some digital memory device so that it stores a picture of the objects. But search wherever you like in the memory, you won't find any actual image, just tiny transistors switched on and off in some pattern, which tell a computer how to produce a visual image on a screen. Similarly there is no literal image in your brain, just a pattern, in the states and connections of its individual neurons, that encodes information from the signals received by your brain from the objects around it. What we know of the brain tells us that some such process must take place; and how else could we be aware of distant objects except by their effects on our brains?

The digital camera metaphor is flawed too. First, there is no one organized place in your brain where the world at any time is represented in the way that an image may be represented in a memory card. Experiments on various sensory illusions indicate that the brain contains overlapping, constantly shifting representations, with the representations of the different senses out of sync. And second, the experiences that you have are not the result of the brain decoding the representations as a computer reads a memory device. Instead, having the representations—the neurons being in a certain state—is what it takes to experience the world that way. Unlike the data in the camera, no additional 'decoding' by the brain is needed.

No matter, what follows does not take account of these two details, and the digital camera metaphor is a good enough approximation for our purposes; the full story, faithful to all we know about the brain, will be along essentially the same lines.

Consider how these representations end up in our brains. To simplify things, let's stick to light signals (the story for sound is in substance the same). Most of the objects that we see, and essentially all of those that we pay attention to in our moment-to-moment interactions with the world (forks, balls, cars, people, etc), are within 100 m or so. Most of them are opaque and prevent us from seeing most more distant objects; additionally, objects further away are unlikely to have any immediate effect on us and so we generally don't pay too much attention to them.

Moreover, my worldline is extremely narrow with respect to the speed of light, so all the light reflected from bodies to me is approximately arriving at a point. Also, light travels 100 m far faster than any change we can register, so the light that reaches my point in space at any instant left each of its sources essentially simultaneously. That is, the light reaching me (from bodies that I'm interested in) at any instant comes from an instantaneous region of space.

Once the light reaches me, it enters my eyes and impinges on their retinas, producing electrical signals that are transmitted down the optic nerve. These signals are distributed to different regions of the brain for different kinds of processing: to detect motion, evaluate distance or relative size, or look for familiar faces, for example. Then the processed information is sent on to other regions for increasingly high-level processing, the effects of which spread through the brain. According to experiment, all this takes place in something like a half to one second, the result being a representation in the areas of my brain relevant to conscious experience. (See figure 11.1.)

Since it is caused by things in space (effectively) at an instant, it is not surprising that the representation is, broadly speaking, of those things

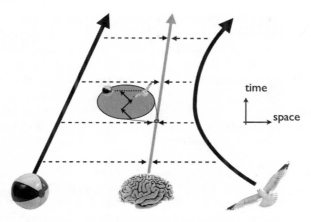

Figure 11.1 Light travels to me from the various things around me—here a beach ball and seagull—and then is processed by the brain to form a representation of the world (see the inset). Since light travels so fast, the light reaching me at any time left the ball and gull simultaneously. Thus the representation is of the arrangement of the ball and gull in space at an instant.

in space (more or less) at an instant. However, as we'll see later, it is important to note that this fact depends on how the brain processes the information.

Of course the process is continual; each representation is quickly succeeded by another. So to sum up, the brain's worldline can be thought of as a long, narrow tube comprised of a series of 'thin' slices, each containing a representation of the world at a moment. (As we have noted, significant idealizations are involved in this picture, but it is faithful to the features we need.)

11.2 EXPERIENCE OF SPACE VERSUS TIME

This description is very suggestive. The representations are surely what is relevant to our experiences: they are in the brain, ordered as our experiences, and represent the things found in our experiences: arrangements of objects in space (not time). But we cannot stop here, because the brain also contains representations corresponding to objects extended in time. Why don't they correspond to experiences of temporally extended things?

For instance, suppose I watch a person for a minute, so that a series of representations forms in my brain over that time, each of the person at a different time. Then the 'thick' (in time) slice of my brain's worldline that coincides with that minute contains that series of 'thin' (momentary) representations. But that series of representations is itself a representation of the person *extended in time*, with a pair of legs at each moment. (Analogously, a strip of movie film contains a representation of a scene extended in time, as a series of images of the scene at different times.) Why doesn't that thick representation correspond to an experience of the person as a millipede?

An explanation of why such a thick representation cannot correspond to an experience, while the thin representations that make it up can, would be an explanation of why our experiences are not of things extended in space *and* time–of why our experiences are non-Tralfamadorian. (A full explanation would also explain how representations in the brain correspond to conscious experiences at all. That problem is too hard to deal with here, and perhaps just too hard. We simply aim to show that only the representations described in the previous section *could* correspond to experiences.)

Have we found another fundamental asymmetry between time and space: that experiences only ever arise from representations at an instant? (That couldn't be exactly right. The representations are thin, not flat: brief not instantaneous.) I'd like to resist that conclusion, because it would add something to the block that can't be understood through physics, something that, moreover, makes the connection between the mind and the brain fundamentally inexplicable. It is preferable to understand features of the mind in physical terms if possible.

Think again of the process by which the brain produces representations in response to external signals: light falling on the retina, for instance. After initial processing, processed information is sent to the different regions of the brain responsible for higher-level processing, which 'confer' with one another, to build collectively a coherent story of what the external signals show. In other words, each representation in the brain's worldline results from a complex 'negotiation' between different parts of the brain. (See Further Readings for an important qualification.) I suggest (not originally) that this fact is essential to their corresponding to our experiences.

In the first place, the coherence of our experiences requires that the parts of the brain corresponding to them are coordinated, each telling a compatible story. And that will happen only if those parts of the brain have worked out the story together. Second, it explains why, for instance, your brain and mine do not produce a single series of experiences but rather two. The parts of my brain interact to produce my experiences, and the parts of yours interact to produce yours, but our two brains just don't interact (at least not to anything like the same degree). Finally, such processes of interaction are the basis for what science understands of the neurological basis of consciousness; they are how we explain the workings of the mechanisms behind experience.

Now, any thick slice of the brain's worldline will include parts of the brain at different times, each containing a representation produced by a different negotiation, based on inputs at different times. So, according to the proposal, such thick slices are not candidates for representations corresponding to conscious experiences. As a matter of fact about the way the brain works, the result of a negotiation is always one of those thin representations we identified, one of objects arranged only in space. So only those can correspond to experience, and we can have only non-Tralfamadorian experiences.

An asymmetry between time and space, implicit in talk of the 'result' of the 'negotiation', is doing work here; it implies an evolution in time, not space. So giving a physical basis to the asymmetry of experience requires giving a physical basis to the asymmetry of negotiation. But that comes down to the physical processes corresponding to the negotiation—they evolve in time.

Picture the brain's worldline, with time quantified using the speed of neural signals, and one of its representations: a long narrow tube and a thin horizontal slice. The negotiation producing the representation looks like a series of back and forth neural communications, starting when light arrives and ending on the slice; see the series of zigzags back and forth in figure 11.1. (Since the slowest neural signals travel at 1 m/s, in the half to one second of negotiation, signals can go back and forth several times if the relevant regions are, say, 10 cm apart.) Because of the way that the brain works, and the way that effects propagate, the negotiation between the parts ends with the representation which we previously saw represents regions of space at (more or less) an instant.

Before we move on to consider the meaning of the asymmetry of those neural processes, I want to make two comments about what does and does not follow from the idea that only the results of negotiations can be experiences.

First, there's nothing obvious in the idea to show that there couldn't be a 'brain' so constructed that the result of each negotiation is a representation of things extended in space *and time*. Our brains function in such a way that the goal of each negotiation is a representation of the causes of signals arriving at an instant (more or less). But in principle one can image a 'brain' built so that the goal was instead a representation of the causes of signals arriving over an extended period, the previous 10 s, say. Such a representation would be something like a long-exposure photographic image; indeed, such a picture of a moving person looks a bit like a millipede. So I don't conclude that *nothing* could conceivably have Tralfamadorian experiences.

Second, these considerations do suggest that the representations corresponding to experience can't themselves be thick, extended in time, however a brain is constructed. Negotiation is continuous, so a part of the brain can't keep representing the same thing over a period (or delay 'ratifying' a negotiation) because it is already busy negotiating the next representation. That is, it seems plausible that experiences have to occur at a time, whether or not they are of extended periods.

11.3 ANOTHER ARROW

Crucial to the explanation of the non-Tralfamadorian, asymmetry of experience is that neural signals propagate *forward* in time (though in any direction in space). Thus the explanation depends on this space–time asymmetry (and facts about our brains). The direction of time was discussed in chapter 10 with respect to the arrow of the weak nuclear force. Although the weak force acts in the atoms of the brain, it is surely not relevant to an understanding of neural processes; they are essentially due to electromagnetic forces, which are time symmetric.

Thus we have reached an important but difficult issue in block time: how to understand a new 'arrow of time', that effects always come after their causes. (Not just by definition: it's not as if sometimes windows break before being struck by bricks and sometimes after and we always call the first event the cause!) The problem is part of the more general issue of why everyday processes are irreversible, having a characteristic direction in time (and not space): colliding pool balls heat up and slow down, tea cools down, plants grow and then wither, and we all age as time goes on. It's implausible that the weak force is responsible for these arrows, because they are largely independent of what happens at the subatomic level. And they can't be understood in terms of the other fundamental laws alone, because they are reversible. So how are they to be understood?

Perhaps (and this is the most popular proposal) the arrow comes from how *things* are arranged—specifically, that they were in some special way much earlier in time.

That way of thinking helps us understand the arrow of specific processes: imagine a room-temperature glass of water at the atomic level, as a collection of fairly quickly moving molecules (ions really). As they bounce around off each other, the molecules exchange energy, so at different times some will move a bit faster and some a bit slower than average, but we expect the water to stay at a constant temperature. The process has no direction in time, since the water is at room temperature at the beginning, middle, and end.

We certainly don't expect there to be so many collisions in which one molecule transfers so much energy to another that the slowed-down molecules join together and spontaneously form ice while the rest of the water, composed of the rapidly moving molecules, heats up. A glass of hot water with ice in it is in a very 'special' state. Moreover it is one that begins a directed process: later the ice will be partially melted and the water cooler, and in the end the whole thing will be at room temperature, and, as we said, it won't go back.

It seems that if *all* occurrences of such directed processes, anywhere and anytime, are understood as the result of such special states, then it must be that the universe as a whole started in some special state. So the proposal is that that special state of the universe is sufficient to explain the causal arrow in individual processes—not just that the cause is the event closer in time to the special state than the effect, but also why the causal processes always point in the same temporal direction. Such a solution would clearly be compatible with the block, but there is plenty of controversy about how to make it work.

Perhaps instead the block has an additional arrow like that of the weak interaction, but one not connected to the fundamental laws in the same way. This one would have to make itself felt in certain circumstances, but not simply because the laws held; the fundamental laws hold for everyday processes, but according to this proposal, that doesn't suffice to make us understand their directedness. We already know that an arrow can be consistently added to the block. However, we would not yet understand how this one is integrated into the physics of time.

Our investigation of the space–time asymmetry of experience ends here, not with an account of the arrow of causation, but with only an indication of how the issue may be developed in the block. That said, we have made considerable progress in understanding how to say something about the striking character of our experiences of time, an issue that seems intractable at first glance. And what we have said has not required attributing to time more than its physical properties.

Indeed, the ideas that we have developed also help us see how the physics of time sheds further light on the other issue of temporal experience that we discussed: its dynamism. For we have now seen why our experiences correspond to a series of representations at a time (of things arranged in space). Moreover, the representations occur in the context of memories, which, because of the arrow of causation, can be caused only by *earlier* representations, and so are only of earlier experiences. And so our experiences are always of the present, *as a moment in a series of earlier experiences*. (Of course they are; the point is that the arrow of causation is relevant to making it so.) But that description is surely another part of what we mean when we talk of the experience of time 'passing'; once again, the physical nature of the brain affects the nature of our experiences.

11.4 PHYSICS AND THE PHILOSOPHY OF PERCEPTION

Physics and philosophy have started interacting in new ways in the last two chapters. Motivating our inquiries were the differences between our experiences of time and space. In this chapter we turned to this issue and explained how some rather basic physical principles could help us understand the differences. The new kind of role that physics thus plays in our discussion is to show how we—here, our experiences—are constrained and shaped in philosophically interesting ways by the fact that we are physical and live in a physical world.

So far I have assumed only that what you experience is *affected* by the state of your brain. However, as we go on I will generally take a stronger position, for the sake of simplicity (and because I think it is true). I will assume that (1) our brain is a physical object, whose behavior is determined by the laws of physics; (2) what we are thinking at any time is completely determined by the way that our brains are at that time: any two times that a person's brain is in exactly the same state, he is having exactly the same thoughts.

This kind of view—'physicalism'—is held by most contemporary philosophers and is opposed to the idea that the mind is some 'superphysical' object over and above the brain, or that thoughts are not determined by the brain (if you think that, then what is a brain for?). This issue is hugely important in the history of philosophy but will not be considered in detail here (see the further readings); I will simply take the physicalist side.

It does *not* follow that my conclusions are irrelevant to the antiphysicalist. When I show that some judgment (e.g., time flows) is possible for a purely physical mind, I also show that it is possible for a superphysical mind; such a mind need not use any of its superphysical powers to make the judgment. The anti-physicalist should be grateful that I have saved him the bother of having to provide a detailed theory of the superphysical mind; the physical properties of the mind—which we all agree it has—get the job done.

Further Readings

Slaughterhouse Five or the Children's Crusade: A Duty Dance with Death (1969) is published by Dell (the quotation appears on 87). For further discussion of the perception of time, there is Daniel Dennett's *Consciousness Explained* (Back Bay Books, 1992). (*Important qualification*: when I speak of a representation as the endpoint of a negotiation, I don't intend to deny Dennett's 'multiple drafts' model. Whatever the truth here, I have just used what I hope is a harmless simplification for our purposes.) I have also been influenced by Jeremy Butterfield's excellent paper 'Seeing the Present', published in *Mind* in 1984.

Richard Feynman in his *Character of Physical Law* explains how one might explain irreversibility in terms of 'special' initial states. For provocative skepticism that this proposal suffices to explain all the arrows of physics, see chapter 4 of Tim Maudlin's *The Metaphysics Within Physics* (Oxford University Press, 2007).

The Mind's I: Fantasies and Reflections on Self and Soul (Basic Books, 2001) is a very nice collection of essays and discussions from a variety of perspectives concerning the place of the mind in a physical universe. It is edited by Douglas R. Hofstadter and Daniel C. Dennett. Another fascinating, provocative (though problematic) book on physics and the mind is Roger Penrose's *The Emperor's New Mind* (Oxford University Press, 1989).

12

Time Travel

Is time travel possible? Of course 'yes', in the sense that we are constantly traveling from past times to future ones, a day at a time. But the interesting question is whether it is possible to move backward or forward through time with something like the freedom with which we can move around in space.

Relativity tells us that in regard to the future the answer is again 'yes'. If my son Kai took a suitable round trip away from the Earth, then according to relativity theory he would 'accelerate' his journey through time, returning more years in the future than he ages. Thus he and his twin brother Ivor could have a joint party celebrating their tenth and twentieth birthdays respectively! (We'll see in chapter 15 how this scenario arises.)

Accepting the possibility of such travel to the future, we can ask the remaining important question this way: Is it possible to move *backward* in time? Specifically, when I ask in this chapter whether 'time travel' is possible, what I want primarily to know is whether it is possible to travel from some moment and place to some earlier moment in time.

12.1 WHAT IS TIME TRAVEL?

Whether you think that the block universe gives a complete picture of time or not, this question is most clearly studied in the block. We have already pictured the worldlines of objects leading from past to present; what might the path of a time traveling object look like in the block universe?

As an illustration, imagine the following scenario: every year, starting at the moment of the summer solstice and lasting for one minute, a 'time portal' opens between a pair of the standing stones of Stonehenge, so that anyone who steps through emerges at the same place but exactly one day earlier. To be clear, we'll say that the portal has two openings, an *earlier* one the day before the solstice, and a *later* one on the solstice, so that someone can travel backward through time by entering the later opening and emerging from the earlier one. (Forward time travel through

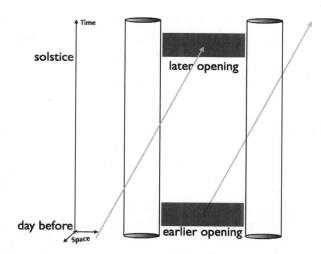

Figure 12.1 A path through the portal that leads back to the past (the times are not to scale). The shaded surfaces represent the openings and are extended in space, between the stones, and in time for one minute. Anything that passes through the later opening does not reappear at the far side a moment later, but a day earlier, as the indicated trajectory shows: it could, perhaps, be the path of a rock thrown through the portal.

the portal involves a 'jump' in time, so it is a bit different from what we'll see in relativity.)

We can picture the situation in our simplified, three-dimensional block universe, in which we pretend that space has only two dimensions (bearing in mind that the three-dimensional block is just a convenient picture to help us think about the four-dimensional universe we actually inhabit). The worldlines of two two-dimensional stones, between which a portal appears, are easily pictured, and the two openings of the portal are two-dimensional surfaces swept out by a line between the stones for one minute starting at the solstice and one day before. Entering the portal means crossing the line during an opening. All this is illustrated in figure 12.1.

But how can we understand the fact that anything entering the later opening will reemerge from the earlier one—that we have a time portal? Recall our discussion of topology in chapter 4. We saw how a cylinder can be thought of as a flat sheet whose edges are identified, so that anything leaving one edge immediately reappears on the other side. In the same way, we have a time portal because the two openings are identified, so that anything, for example a rock, passing through the later one immediately appears through the earlier one. In figure 12.1 the line between the two stones is identified with the line the day previously for one minute, hence the point where the indicated path enters the later opening is the same point as it emerges from the earlier opening. In a block with three spatial dimensions and one time dimension, the opening between two stones is a two-dimensional area, so a portal means two surfaces are identified.

(A couple of qualifications: first, we not only identify the earlier and later lines, we need to disconnect them from the space behind them, so objects do not emerge from the back of the later portal as well as the earlier one. Second, someone entering the earlier opening will emerge through the later opening because of the identification, just as one can travel either way around a cylinder.)

Let's just be clear about what is entailed here. If something is around a day before the solstice, then passes through the portal, then it will be around *twice over*. Indeed, this is shown in figure 12.1: between the solstice and a day earlier, two stages of the path are present. Thus if *you* travel through a portal, you could meet yourself. That would be strange, but not logically impossible, for you would be meeting yourself at an earlier age, a day younger in our example. It would be rather like meeting a clone of yourself who was created a day after you and who had lived a very similar life to you.

Finding a person at different stages of her life existing at one time is exactly the kind of occurrence we expect if earlier and later parts of the block are joined to make a portal. But it is quite different from what happens in movies in which someone gets to live over some period of his life again: *Groundhog Day* is an obvious example. That is not going back to some part of the block, because the block doesn't change; if you travel to the past in the block, it is just the way it always was (so it always contained you coming out of a time portal!). In *Groundhog Day* something else is going on—travel to *another* universe much like ours, perhaps.

12.2 IS TIME TRAVEL POSSIBLE?

If such a portal can exist, then time travel is apparently possible, but can a portal exist? What makes the discussion of time travel so interesting is that—as we shall briefly see in chapter 15—according to Einstein's general theory of relativity, paths to the past are indeed possible. (I don't mean to give the impression that time travel is impossible in nonrelativistic physics: the point is that the laws of Newtonian physics never entail the existence of a path that would bring a body into its past, while the laws of general relativity actually entail that in some situations such paths must exist.)

The 'portals' of general relativity are not as simple as the one we're considering, and for various reasons they might be impossible to create or use—certainly far, far beyond our present abilities. One might even argue, from the fact that we haven't met any time travelers from our future, that time travel is at the very least practically impossible. But that would be a hasty conclusion: it might be that in the future beings will have the technology to build time portals but be unable to place the earlier opening just anywhere they want.

For instance, one kind of portal, proposed by Kip Thorne (see Further Readings), involves a wormhole in space—essentially a shortcut from one place to another. Suppose Kai takes one end of the wormhole on his relativistic round-trip, while Ivor keeps the other end with him on Earth. Since Kai's trip takes him 10 years into Ivor's future, it also takes Kai's end of the wormhole 10 years into Ivor's future, making a shortcut from one time to another: a time portal! But the earlier opening of such a portal can only exist *after* Kai's departure; before that the wormhole connects simultaneous locations. So beings with the technology to construct such a portal would not be able to travel back arbitrarily far in time, to the present, for instance. Thus, the absence of visitors from the future is no argument against time travel and the intrinsic interest of the topic makes it well worth pursuing. (Michael Lockwood makes the same point in his book *The Labyrinth of Time*, which is again recommended; see Further Readings in chapter 10.)

More than sensationalism motivates a discussion of time travel. As we discussed in chapter 1, most physicists believe that general relativity must be replaced by a quantum theory of gravity, perhaps string theory. New theories are usually sought because existing theories are found to have flaws, and knowing at least some of those flaws is very useful in figuring out what a better story might be. But rather little is known about the flaws of general relativity, because its predictions—those that we know how to derive from it and that can be tested—are hugely reliable.

As we will see, the possibility of time travel causes a number of potential paradoxes, so it may be that permitting time travel is a failure of general relativity. Many physicists think that this may be the case and have hence studied time travel in general relativity, looking for clues to a quantum theory of gravity. (Indeed, some have argued that all time travel may be prohibited by a quantum theory of gravity; see the Further Readings.) The details of this work lie outside the scope of this discussion, but what we will be able to see is how things might go wrong, and alternately how time travel might be accommodated by physics.

12.3 THE PROBLEM WITH TIME TRAVEL

Consider then the following story, which the existence of a portal would seem to permit. I drive to Stonehenge, eager to time travel, elbow my way to the stones through crowds of druids and hippies, and as the solstice begins, leap into the portal and come out a day earlier, sick to my stomach—time travel apparently makes me extremely nauseated. I am so upset by the experience that I wait a day until I see myself (my one day *younger* self) coming through the crowd, accost myself, explain the consequence of time travel, and talk myself out of entering the portal. By not time traveling, I avoid getting time-sick and of course I never get to stop my earlier self.

Every step in this story seems quite plausible, but the complete story is quite impossible, because in it I do and don't enter the portal, I do and don't get time-sick, and I do and don't stop myself, which is contradictory! (This reasoning works at least on the obvious view of time, in which there is only one present at any moment. Science fiction sometimes allows time to 'branch' into different 'time lines' so that there can be many different, incompatible simultaneous presents, but we won't pursue such an idea: we want to know whether time travel is possible in the more familiar kind of nonbranching time.) The problem with time travel is that it seems to allow logically impossible situations, a situation that in turn seems to imply that time travel itself is impossible.

But that conclusion is too hasty, because not all stories involving time travel are contradictory. Suppose that things happen as before except that I fail to stop my earlier self from entering the portal, perhaps because I am not sufficiently persuasive, or because I don't get time-sick and so don't try to stop myself. Then we have a story that is perfectly logically consistent. Thus time portals and time travel don't necessarily lead to contradiction and so are not logically impossible.

However, we live in a world with specific physical laws, and so the real question that we should address is whether time travel is consistent with what we know of the laws of physics; is it physically possible? We can put the question this way: consider the *physical state* of the universe— the arrangement and properties (including motions, masses, charges and so on) of all material things—before the earlier opening of a portal. Is there *at least one* such state whose future evolution is consistent with the presence of a portal in its future, a state that is the start of a consistent time travel story? (Or to draw on an earlier analogy: is there one volume in the library of books allowed by the laws of physics that involves a time portal?)

If time travel is physically possible in this sense, we can ask whether or not it requires a special state of affairs. Are *all* physical states compatible with time portals in their futures? Do they all evolve to give consistent time travel stories according to the laws of physics? (Or in our analogy, for every possible opening sentence, is there a volume of the imaginary library that starts with that sentence and which involves a portal?)

Since it is physics that tell us what physical states there are and how they evolve, the answers to these questions clearly depend on what the correct physics of the world is. We'd like to answer them for general relativity, since that is currently the best understood, most physically accurate theory of the universe (setting aside for now the complexities of the very tiny things treated by quantum mechanics). General relativity allows time travel in principle; but general relativity permits a huge number of possible universes, most very different from ours, and so 'in principle' possibility doesn't tell us whether time travel is allowed in our universe, or only in universes that are very different. Whether any or all of the possible states of our universe are compatible with time travel is a very difficult issue indeed, but we can make some progress by distinguishing two kinds of question.

On the one hand, could the kinds of things we find in our universe be arranged in such a way that they would produce a portal according to the laws of general relativity? For example, if matter were arranged as a rotating dust cloud that completely filled the universe, then every point would be a portal (of somewhat different kind from ours). Matter is not actually arranged that way, but is it arranged so that some other portal could arise, either naturally or artificially? The strictly logical consequences of the theory and what we are firmly justified in assuming about the universe give a tentative 'yes' (which many physicists take as evidence of the failure of current knowledge).

On the other hand, now assuming the existence of such a portal, we can ask: according to general relativity, do *any* possible initial states of our universe lead to consistent time travel stories involving a portal? If so, do *all* such initial states lead to consistent time travel stories? (By 'initial state' I don't necessarily mean the state in which the universe came into being, but a state some time before the portal; it is 'initial' in the sense that we are interested in what comes after.)

These are difficult questions, which is why physicists have found answers only in some very simple cases, themselves more complicated than is suitable for our discussion. We shall pose them for even simpler, artificial (or 'toy') physical theories that help us understand the current understanding of time travel in general relativity. (One of the tricks to physics or philosophy is to use good toy theories that strip all the unnecessary complications from a complex situation.) In the next chapter we shall consider whether time travel really makes sense, physics notwithstanding—whether physical theories, such as general relativity, that permit time travel make sense.

12.4 POSSIBLE AND IMPOSSIBLE TIME TRAVEL

Here's an example of how a toy physical theory might run into problems with time travel, in a way that is similar to the earlier contradictory story involving me. Imagine that the only thing in the universe is a light bulb with a power source and light detector, subject to the 'law' that if it sees a bulb that is turned on, then it switches itself off and stays that way permanently, and if it sees a bulb that is off, then it turns itself on and stays that way. Suppose further that there is a time portal so that the bulb can pass to its own past.

If the bulb is on when it enters the portal, then its 'younger' self will see it and turn itself off, and it will enter the portal switched off; similarly, if it is off when it enters the portal, then it will switch itself on and enter

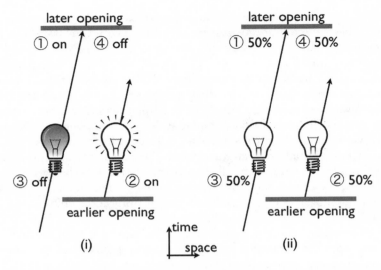

Figure 12.2 (i) When the bulb first detects another bulb it turns itself off if the other is on and vice versa, then stays that way. It is thus impossible for the bulb to travel through a portal. For instance, if (1) the bulb enters the later opening on, then (2) it emerges on. Thus (3) the bulb on its way to the portal will turn itself off, and (4) enter the portal off—in contradiction with (1). (The same contradiction arises if the bulb enters off.) (ii) But if the bulb merely inverts the brightness of the bulb it sees, then it can time travel, as long as it is at 50% illumination as (1), it enters the portal. For then (2) it emerges at 50% and so (3) sets itself to 50% on its way to the portal, (4) entering at 50%, consistent with (1).

on after all. Thus if the lamp enters the portal turned on, then it must enter it turned off, and if it enters it turned off, it must enter it turned on (see figure 12.2[i]).

And so, given these laws, there is no consistent story in which the lamp passes through the portal, no possible story in which the lamp time travels. The problem is that the portal allows the lamp to act on its own past, destroying its current state. If our physics were like the physics of this world, then time travel would be physically impossible; there would be no physical state leading to a consistent time travel story.

Now modify the physics so that the bulb need not be either completely on or off but can have any level of illumination in between, and so that when it sees another light bulb it 'inverts' the brightness, making itself as bright as the other is dim. Then there is a possible story involving time travel (see figure 12.2[ii]). As long as the bulb going in is exactly 'half' on, its effect on its earlier self will be to turn it half on before it enters the portal; half on inverted is half off, which is of course also half on. (There are no other possible time travel stories in this case; if, say, the bulb enters the portal at more than 50 percent illumination, then it will set its 'younger' self to less than 50 percent illumination and so enter the portal at less than 50 percent, which of course produces a contradiction.)

This model illustrates how there can be a positive answer to the question of whether physics allows any consistent stories: here there is a way for the future to act on the past that is not self-destructive. In general terms, there will be a consistent story if physics allows an interaction that 'duplicates' the state of the time traveling system in its earlier self—half on, for instance.

As far as physicists can tell from studying very simple systems, according to general relativity if our universe permits portals then there are initial states that lead to consistent time travel stories. Our universe is like the second world rather than the first.

One kind of example that has been studied involves a single pool ball that is deflected into the portal by its later self emerging from the earlier opening. Wherever the portal openings are placed, there is always some initial trajectory of the ball (before the collision) such that it is knocked by its later self into the portal in just the right way to emerge from the earlier opening on just the right trajectory to cause the deflection. (A problem in fact tends to be that the initial state does not uniquely determine the outcome but is compatible with many self-collisions.)

Thus we can tentatively say, given what is known about the world, that physics allows time travel without paradox, that time travel is physically possible, which is of course a long, long way from saying that it is technologically possible. Moreover, to say that some possible initial states produce consistent stories is not to say that all of them do, and if not, then we will need to say something about how the states that are pregnant with paradox are avoided. To understand these points, consider another toy theory.

Imagine a world in which the physical laws describe a new kind of cellular automata. Suppose that space is one-dimensional and discrete, made of cells that can be either occupied or empty; any distribution of occupied and empty cells—any state—is possible. Consider any cell and the two adjacent cells; call these *three* cells the 'neighborhood' of the cell. Let time be similarly divided into discrete moments, one second apart. Then this is the law of the theory: take each cell and its neighborhood at some moment in time; at the next moment the cell will be occupied if exactly one or exactly two of the cells in its neighborhood are occupied now; otherwise it will be empty. See figure 12.3(i) and (ii), for instance.

Now suppose that the 'cellular' universe in question has a time portal that transports the contents of some cell at a particular moment back to a cell five seconds earlier. Thus if the later opening is an occupied cell, then that state is transported back to the earlier opening, and if unoccupied, then that state is transported. That is, the openings are identified; the 'two' cells are really one and so must have the same state. (Similarly, since the two openings of the Stonehenge portal are identified, they must be in the same state: containing me, perhaps!)

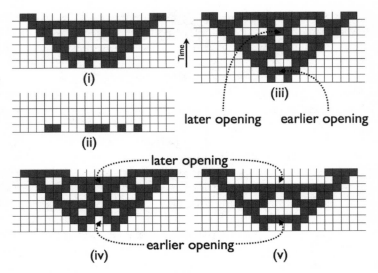

Figure 12.3 (i) The evolution of a pattern of four occupied cells (the bottom row), for the six following moments, according to the given law. (ii) is left blank after the first generation for you to fill in! (iii) An evolution with a portal; the two cells indicated are identified. The evolution pictured is possible: every cell is as it should be according to the laws and the two ends of the portal are the same. The initial state in (iv) and (v) is incompatible with a portal in the location shown. If (iv) the earlier opening is unoccupied, then so is the later (since they are the same cell), which is inconsistent with the law and the state one second earlier. If (v) the earlier opening is occupied, then so is the later, again inconsistent with the law.

Figure 12.3(iii) shows that, as for the second light bulb and for general relativity, there is at least one initial state that leads to a consistent story involving the portal; if the only occupied cells one second before the portal first opens are the two adjacent to the portal cell, then there is an evolution according to the laws in which both openings are occupied. If you like, given that state, the transportation of the occupied later opening to the earlier opening leads, according to the laws, to the later openings being occupied; what comes out of the portal leads to the right thing going into the portal. And so time travel is physically possible in this world.

However, figure 12.3(iv)–(v) shows that not every possible initial state leads to such a happy outcome. Suppose that the only occupied cells one second before the first opening are those two cells away from the portal. Then, as you can see, (iv) if the earlier opening is unoccupied, then according to the laws the later one must be occupied, but then the portal means that the earlier opening is occupied after all; but (v) if the earlier opening is occupied, then the laws mean that the later one is empty, and so the earlier one must be too. Either way we have a contradiction: the earlier opening cannot be both occupied and unoccupied, since no cell can be both at the same time; so since (iv) and (v) are the only possibilities—the opening must be either occupied or unoccupied—there is no consistent story with this initial state.

And so in this model, while time travel is physically possible, the occurrence of a portal *in the future* restricts what states can occur now. At a given time, absent a future portal, any of the cells can be occupied and any can be unoccupied, but the occurrence of the portal considered means that one second before its earlier opening, certain states—the one in (iv), for example—are impossible, because they would lead to logically impossible stories; of course any prior state that would evolve into such an impossible state is also impossible.

It's as if in the Stonehenge story it turned out that inconsistency would result: I would time-travel and prevent myself from time traveling, if, say, I wore a red T-shirt two days before the solstice. Then it would be impossible for me to be in a state two days earlier that would take me through the portal and in which I was wearing a red T-shirt. And let's be clear, it is the future presence of the portal that makes the state impossible. I could be in the very same state, wearing a red T-shirt, if there were no portal; I would simply pass through the two stones on the solstice. Correspondingly, the state in (iv)–(v) is perfectly possible if there is no portal in its future.

I have no reason to think that wearing a red T-shirt would, given the laws of physics, lead to a contradiction with a future portal; in fact it seems unlikely that it would. The point is to imagine a situation in which some state that otherwise is perfectly possible somehow becomes impossible if it has a portal in its future, and how counterintuitive that situation seems. However, as I will explain in the next chapter, while such a situation would be odd, I do not think that it would make time travel incoherent in some way.

For now we can see the importance of the question of whether the physics of our world—of general relativity—is like that of the cellular automata world, or whether every possible initial state is compatible with a portal in its future. This question also has been studied only in simple cases. In those cases the answer seems to be that there is a consistent future for any initial state with a portal in its future: matter can be distributed freely without fear of contradiction. As far as we know, then, if general relativity permits time travel—and in our universe that is a big 'if'—then it requires no special restrictions on the states of the world.

12.5 THE PHILOSOPHY AND PHYSICS OF TIME TRAVEL

In this chapter we have made use of some earlier ideas in order to understand what time travel might involve: if regions of a block universe at different times are topologically identified, then trips to the past (and future) are possible. We have also used physics to pose and answer the question of whether time travel is possible. That is, we ask whether there are any states that lead to consistent futures, and we use the laws of physics to find a precise answer. If that answer is 'yes', then we can also ask whether all states are consistent with future portals. Once again we see how physical theories can be used to address philosophical questions, asking whether the laws of the theory have some suitable mathematical feature and then proving mathematically either that they do or that they don't.

The answer is not really known for our best theories, such as general relativity, though in simple cases all states do seem compatible with portals. (Even better, general relativity allows portals, perhaps even in our universe.) We have seen, however, a very simple toy theory in which we have been able to answer the question. That is, in the world of cellular automata which we explored, we saw that there are some states in which a portal is possible and some in which it is not.

The toy model was not very realistic, of course, but it was very instructive in a number of ways. First it showed how toy models can help us to understand a question and to see how it might be answered in a manageable case. Sometimes toy models help physicists and philosophers make a first step to a fully realistic solution, though here the main point was to help nonspecialists understand the problem. But we honestly posed and answered the question for our theory, showing quite accurately what philosophical work in the foundations of physics often looks like.

The model also gives a good picture of the situation facing us, showing quite clearly how future portals might be relevant to the current state: if there is one in the future, then that fact may bear on what states are possible *now*. That idea sounds strange, since we think the past can affect the future, not vice versa. And even with a time portal, surely the future can affect the past only *from the time of the earlier opening on*. Here the constraint on states would apply even earlier.

However, as we shall see in the next chapter, that strange conclusion is in fact a key to understanding the paradoxes of time travel.

Further Readings

There are two recent books that give popular accounts of the current state of play regarding the physics of time travel: *Time Travel in Einstein's Universe: The Physical Possibilities of Travel through Time* by J. Richard Gott (Mariner Books, 2002) and Paul Davis's *How to Build a Time Machine* (Penguin Books, 2003).

You might also like to read Carl Sagan's *Contact* (Simon & Schuster, 1985), which uses a kind of wormhole allowed by general relativity, that could be used for time travel. This, and much more about time travel and other issues in general relativity are discussed in the physicist Kip

Thorne's wonderful *Black Holes and Time Warps: Einstein's Outrageous Legacy* (W. W. Norton, 1995).

I have also been heavily influenced by an essay by Frank Arntzenius and Tim Maudlin on 'Time Travel and Modern Physics' on the *Stanford Encyclopedia of Philosophy*: http://plato.stanford.edu. The *Encyclopedia* is a great resource for anyone wishing to find out more about specific topics in philosophy from some of the best people currently in the field. (I should say that the essay by Arntzenius and Maudlin is probably pitched at a higher level than most of the entries.)

13

Why Can't I Stop My Younger Self from Time Traveling?

We've seen how time travel might be possible according to physics, but an important difficulty remains. Unless it can be defused, then time travel will turn out after all to be paradoxical or incoherent, and consequently so will any physics that allows it, including general relativity, as things stand.

The problem is this: if a time portal exists, then so does the potential for an impossible story, one in which I travel back through time and stop myself from traveling through time, for instance. Such a story cannot happen because it is logically inconsistent, but given the existence of a portal, there's nothing we would take to be impossible about any individual part of it: I can enter the portal; I can wait for my younger self to arrive; and I am surely capable of preventing someone just like me (and hence in principle my earlier self) from entering the portal, especially given a day to prepare. But since we know that the whole story is impossible, if these steps in the story are not, then it must be the portal and time travel itself that are impossible.

The purpose of this chapter is to understand just why the impossible stories are impossible, not because portals are impossible but because, counterintuitively, it is impossible (in a certain sense) for me to stop my earlier self from entering the portal.

13.1 PHYSICS MIGHT STOP ME...

The first answer, based on the previous discussion, is that physics might prevent impossible stories involving a portal, if the laws are such that every possible state evolves in a consistent way, even in the presence of a portal. As we saw, this is not true of the cellular automata, but it may be true of our universe, given its material constitution and the laws of general relativity.

Then it immediately follows that there is no initial state that would produce the inconsistent story of my stopping my younger self: after all, every state leads to a consistent story. Put another way, pick any state of the world (including me) that, according to the laws, ends up with my passing through the portal and trying to prevent my passage. Since it is a state from which I make it through the portal, and since all possible initial states lead to consistent stories, it simply must be an initial state in which I fail to stop myself. And so if all initial states produce consistent stories, physics itself—what states there are and how they evolve—permits no impossible stories and thus prevents me from stopping my earlier self.

Note that this line of thought assumes that the laws of physics govern the actions of people: that facts about the initial physical state and laws are sufficient to determine whether a story will be consistent, and that there's nothing the people involved can do to change that fact. We relied on a similar idea in chapter 11, and later in this chapter we will discuss further how physics limits the actions of humans.

13.2 ... AND IF NOT, LOGIC WILL

Now suppose instead that physics is such that if there is a portal, some initial states evolve into consistent stories, but not all do. The world of cellular automata works like this, and perhaps ours does too. So suppose, for example, that there is a state of the world (including me) which, according to the laws, seems to evolve in such a way that I both enter a portal and prevent myself from entering it. We saw examples of this kind of thing in the previous chapter.

Such logically inconsistent histories are impossible, so logical consistency makes such a state impossible as an initial state *if* the portal is in its future. That is, any initial state that ends up with my entering the portal must, as a matter of logic, be one of those states in which I fail for some reason to stop my earlier self. That is, *some states that otherwise would be possible are rendered impossible by the presence of a portal in the future.* In this sense it is just logic, not physics (and not the impossibility of time travel) that prevents me from stopping my earlier self.

One might question this line of thought, concerned about how the future occurrence of a portal can 'act' via logic to prevent a certain initial state from occurring now. How can events in the present take account of a possible inconsistency that hasn't yet occurred?

According to the block universe view of time, there is really no worry here: the universe in time and space is a unified whole, and logic demands that the whole be consistent. Thus there just are no possible block universes with portals and problematic initial states, only consistent universes. The demands of logic are satisfied all at once, not just moment by moment as time passes. In our library parable, either the whole book is in or it's out; there's no being in until something goes wrong.

But we don't have to appeal to the block in this way, because a little careful thought will show that the way that the future restricts the past in our time travel story is actually quite familiar.

Suppose I tell you that I recently took a boat trip and got terribly seasick, as usual, and ask you for a story with this as the final event. There are a lot of stories one can tell about the day before my trip, but of course given the boat trip, none of them—on pain of inconsistency—involves someone helpfully reminding me of my seasickness ahead of time and thereby preventing me from taking the trip. And so, even though there are states involving me which evolve into stories in which people do prevent me from taking boat trips, if it is already given that I did take the boat trip, logic ensures that none of them is a possible initial state.

That is, we don't need time travel to see that the future restricts the past; in fact our quite ordinary practice of retrodicting (reasoning about the past) relies on it. And just the same thing is happening in the time travel story; given that I do time-travel or go boating, no state that would guarantee that I didn't take the trip is logically possible.

Our discussion here dispels the worry at the end of the previous chapter: how can the future appearance of a portal make a difference to what states are possible now? The answer is again that it is perfectly ordinary for given facts about the future to affect what is presently possible, whether those facts concern boat trips, time trips, or portals. What makes such restrictions seem weird is that we so rarely know them ahead of time, and so we so rarely take them into account, the way we take the past to constrain the future.

Now the future doesn't have to be the way it will be; there are other possible universes in which things happen differently. And naturally, neither do the given facts rule out anything that doesn't prevent them. Hence there are some pretty good senses in which I *can* stop myself from entering the portal.

On the one hand, we can imagine a story in which there are two portals, one from the solstice to the previous day, and one from the following equinox to a year previously. Then suppose I travel through the equinoctial portal and prevent my younger self from entering the solstitial one; no paradox there. Or, as we shall discuss in the following section, though I cannot stop my younger self from entering the portal I have traveled through, I would be able to stop someone who seemed the same for all intents and purposes.

When we say what 'can' and 'cannot' happen, we always mean given some constraints or other, which we either state explicitly or leave to context. 'You can't go in there' means one thing given that the door is locked and another given that it is a private party. But it is hard to imagine a context in which one would naturally take the assertion that 'I can't stop myself entering the portal' to mean '. . . assuming only that I do enter it'. More likely we would take the constraints to be much weaker—some medical condition perhaps? So it is pretty misleading just to make the bald assertion; however, for brevity I will continue to do so, now that you know what I really mean.

13.3 MY PRECISE PHYSICAL STATE STOPS ME

If time travel is possible, then something has to prevent inconsistencies; that that something is either logic or physics is unavoidable. However, our answer so far lacks the kind of specificity that we normally expect from physical or logical explanations. Physics can prevent me from running faster than a train because of the mechanics of my body, or accelerating to the speed of light because doing so would take an infinite amount of energy. Logic can prevent me from being red all over because I'm green, or from being more than seven feet tall because I'm less than seven feet tall.

But although it is physics or logic that prevents me from stopping my earlier self, we have so far said nothing in similar detail about why this is so. If no such specific answer could be given, then we would just have to accept the explanation we have so far. However, we will be far happier if we can address the very strong intuition that if it is physically or logically impossible for me to stop someone from walking between a pair of stones, then there must be something about me or the world that prevents it.

Let's explore this mystery some more, first by considering some 'possible' variations of the 'impossible' story—stories that are like it in some ways but that are not contradictory.

(i) Consider first a story in which I travel through the portal but somehow don't stop myself: I don't even try, or I give up after failing to persuade my very determined younger self, or, failing to persuade myself, I attempt and fail to restrain myself physically (see figure 13.1). You can imagine other scenarios. As we've seen, either physics or logic mean that any story in which I do travel through the portal must be like this, on pain of contradiction.

The other two stories are more baroque but should be physically possible if this one is. They involve perfect duplicates of me. I don't think that there are such things, but what matters in these examples is that certain things are possible, and I do think a duplicate is possible in principle, if not practically. It is tempting to call the duplicates my 'clones', but properly speaking my clone is merely a genetic copy of me, which could be quite different from me in many ways. The duplicates that I have in mind are exactly like me in every way, down to to the last hair on my head, the last neuron in my brain, and my last subatomic particle;

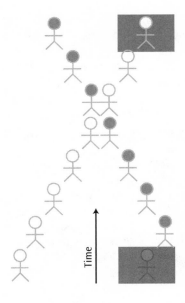

Figure 13.1 I come from the left and enter the later opening (meeting my later post-time-travel self on the way). I emerge from the earlier opening feeling sick (my head is shaded to keep track of my later self). I meet my earlier self headed toward the portal but somehow don't stop myself from traveling through the portal.

hence they are constrained by physics in exactly the same ways that I am.

(ii) In the first story involving a duplicate, initially I am exactly as in (i) and my doppelganger—Mick—is a perfect copy of me. I proceed again to the portal, travel back one day, get sick, and decide to try to stop myself from ever time-traveling.

However, before I meet up with my younger self I run into Mick on his way to the portal, and the result of our meeting is a conversation in which I realize that I will not be able to stop my younger self, and in which Mick decides not to time-travel so that he can avoid time travel sickness; I don't interfere with my younger self, and Mick doesn't try to enter the portal, so this story is manifestly consistent. All this is shown in figure 13.2.

Now suppose that the universe is ultimately a purely physical entity, by which I mean that the physical state and physical laws of the universe completely determine everything about it, including, for example, everything about you: your height, eye color, and more controversially what you are thinking, hoping, and dreading.

You may not agree with this assumption; many people believe that such properties are not fixed by physics but determined by an immaterial mind or soul. However, grant the assumption for now because using it is the clearest way to make some important points; in return I promise to come back and explain how to make similar points for those who are squeamish about my physicalism.

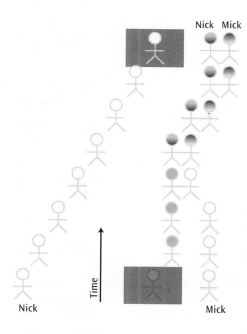

Nick   Mick

Nick   Mick

Figure 13.2 The story starts with me (just as before), an exact physical duplicate, Mick, and my later self emerging from the portal. This time my post-time-travel self meets Mick and persuades him not to time travel.

Let's also suppose, plausibly, that the particular laws governing me and my doppelgangers in the stories are (future history) deterministic: our states at one time determine, according to the laws of physics, unique states at later times. In particular, suppose deterministic laws of physics govern the conversations in the stories, so that their outcomes are completely determined by the initial states of the participants. (Strictly we should take the environment into account as well, but we'll assume that it is fixed and constant, so that the only things that we have to consider varying are the people involved.)

Consider then the conversations that occur in these two stories: in (i) I fail to stop the person I meet (namely, my younger self) from entering the portal, but in (ii) I succeed in stopping him (namely, Mick). The assumption of determinism means that the states of the two parties in (i) must be different from their states in (ii); otherwise they would have to have the same outcome. The difference does not lie in the younger person—myself or Mick on our way to the portal—for they are perfectly alike according to the story. So it must be the person emerging from the earlier opening of the portal (me, that is).

That seems perfectly reasonable: although I start out the same in each story, in (i) I emerge after someone (my older self) tried to stop me whereas in (ii) I enter the portal without hindrance. It seems quite reasonable to suppose that this difference in my past would have an effect on me; perhaps, in (i), experiencing my failed attempt to stop me from entering the portal makes me not quite determined enough to stop myself, whereas in (ii) I retain my determination and so succeed in stopping Mick, even though the people I attempt to stop in each story are absolutely identical.

This example is nice for understanding how physics puts restrictions on the various steps that make up time travel stories, which is what we are after in order to understand something specific about how logic and physics prevent paradox. Things get more complex if immaterial minds and souls are added to the story, but for one way of thinking about these things the same sort of reasoning should apply. That is, if my earlier and later selves in (i) are exactly like Mick and my later self in (ii), down not only to our physical states but also to the states of our minds and souls, then the outcomes of the two conversations ought to end up the same—but they don't. Further, the difference can be traced to my different histories before entering the portal in each story. It's reasonable to think that I will be a different person according to whether or not I have been accosted on my way to the portal, as I have in (i) but not (ii).

Some people won't be happy with even these claims, because they think that they interfere with free will; even if we are exactly alike materially and immaterially, then there should still be the possibility of our choosing different outcomes. There are old and deep and expansive philosophical arguments about this claim, which can be explored through the suggested readings; it shouldn't be much of a surprise to hear that I hold a different conception of free will, which is compatible with physical determinism, though that is another story. More important, the main point that I want to make in this chapter is compatible with any account of free will, as I will explain later.

(iii) In the second story with a duplicate, initially I am slightly different from before in some way: perhaps one or two of my neurons are in different states, so that I'm not quite so determined to enter the portal. My doppelganger, Dick, is a perfect copy of me as I would emerge from the portal in story (i): nauseated and determined to stop me from entering the portal. See figure 13.3.

If I were initially exactly as in (i), then the ensuing conversation between Dick and me would start just as the conversation in (i) and (assuming still that determinism rules the interaction) it would have to finish as before, with my going on to enter the portal. But since I am slightly different, then it is compatible with determinism that Dick successfully dissuades me, and I don't travel through the portal. That is, in this final story Dick meets me on my way to the portal and persuades me not to time-travel; in fact no one time-travels, and so no contradictions occur.

We can use these three stories to address our question; what is it about me in (i) that makes it impossible for me to stop my earlier self from

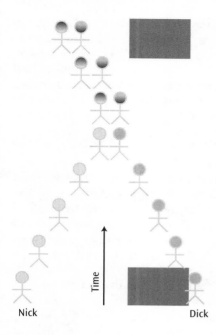

Figure 13.3 In this story I start off a little differently from the first story, and therefore a duplicate of my nauseous post-time-travel self, Dick, can and does stop me from entering the portal.

Nick

Time

Dick

entering the portal? My younger self and I are pretty evenly matched, so how can it be impossible for me to stop him?

Scenario (ii) shows that if I had emerged even slightly differently—because perhaps I had entered the portal without encountering myself—then I could have stopped someone exactly like my earlier self, namely Mick. What (iii) shows is that even emerging exactly as I do in (i), it is still the case that I could have stopped someone almost, but not absolutely, exactly like my younger self.

These examples help dispel a great deal of the mystery, for they show that it is the precise states of me and my younger self that make it impossible for me to stop him, and that is the kind of more specific limitation that we were hoping for. That is, what we have discovered is that it is impossible for someone in my precise state to stop someone in his precise state, but if either of us were slightly different then it would be quite possible.

And that possibility doesn't seem odd after all; when I say that something is possible for me, what I mean is that within the range of precise states in which I typically may find myself, and the range of possible starting states, at least some but not necessarily all of them will lead to one of some desired outcome states. Since we don't know our exact physical state, our assessments of what is possible naturally do not take them into account.

For instance, is it possible for me to walk along a narrow beam? Yes, there are some normal initial states according to which I have sufficient balance and concentration to do it—though not all states, and perhaps not the very state I am in now. Scenarios (ii) and (iii) show that a similar situation must hold in (i): if I had been a bit different, or if I were faced with a slightly different person to dissuade, then I could have done it, and so in the normal sense I am capable of stopping myself, even though I am not in the precise state that I occupy.

13.4 LIVING IN A PHYSICAL UNIVERSE

I've been assuming both that our states are entirely determined by our physical states and that how those states develop is determined by the laws of physics: hence the outcomes of the conversations are determined entirely by our physical states and the laws of physics. On that basis we can conclude that it must be my precise physical state that prevents me from stopping my earlier self, while I retain all my normal powers of stopping people.

Some people may feel queasy about the idea that one's state is exhausted by one's physical state, because they think of the mind or soul as somehow a distinct part of a person. I don't, but even if I'm wrong, exactly the same conclusion must occur if one wants to add an immaterial mind or soul to the picture, if one still accepts that the material plus immaterial state determines the outcome of the conversations. That is, if one of us were even slightly different physically, mentally, or in our souls, then the conversation could end differently, and so in the normal way of speaking, which does not take our exact state into account, I am capable of stopping myself.

Giving up the other assumption as well will, however, change things. Undoubtedly we make judgments about the different degrees to which people act from 'free will' on different occasions: Rosa Parks chose to keep her seat on the bus but acted under duress when she was taken to jail by the police. This idea is important in one's making moral judgments, since we mitigate blame when the ability to choose is diminished. A long philosophical tradition, backed up by a sensible intuition, holds that determinism is incompatible with such free will: if your future actions are determined by your current state (physical and perhaps immaterial) then how can you be said to have any choice in the matter? Aren't the laws of physics pushing you around, regardless of your will?

Like many other philosophers, I think that line of thought is a mistake. For one thing, an old argument points out that if the state of your mind (whether ultimately physical or not) does not determine your actions, then they seem to be *random*, but randomness is not choice. That is, denying determinism does not make 'free will' any easier to understand, and perhaps makes it impossible. But let's suppose for a moment that free will is incompatible with determinism and that we have it (whatever that means), and let's consider briefly where our reasoning stands.

If physics doesn't govern our actions, then in cases involving people we may or may not be able to appeal to physics to understand why I can't travel back in time and stop my earlier self from so doing. It may be that my physical state is such that I am incapable of stopping my earlier self. Whatever the nature of free will, physical constraints regularly prevent me from exercising it: however much I will it, I can't fly just by flapping my arms. Similar physical constraints might rule out my stopping my earlier self. But if free will goes beyond physics, then even if I'm physically capable of the act, I may choose not to. In this case, logic will prevent me from stopping my earlier self; since it is logically impossible to travel and not travel through the portal, given that I do, and given that my will is not thwarted, it's a logical consequence that I don't choose to stop myself. (Of course, for things without free will—billiard balls and machines, for instance—everything said earlier still holds.)

A full discussion of free will would take us far from our main topics (see the Further Readings) so a short discussion will have to suffice. Let's start by noting that there is no reliable experimental evidence that we can will things to happen in conflict with the laws of physics. No evidence shows that you can raise your hand or send signals down through your nerves to the muscles in your arm or produce neural activity in a way that is not perfectly in accord with the laws. To be fair, an experiment to show that *everything* happening in your body is compatible with the laws would be utterly impractical. But the point is that, *absent evidence to the contrary, since the laws of physics hold everywhere that they have been tested, it is reasonable to conclude that they hold of us too.*

What is also clear is that when we talk about free will, we have in mind a person with a range of *options* that they *understand* and are *capable* of undertaking, and that have similar kinds of *costs* (it's not the case that all but one will kill you). Moreover, we expect someone with free will to be able to *evaluate* the actions against their costs and *benefits*, and against a complex collection of *beliefs and values* concerning the world, morality, beauty, theology, and so on. And of course the person is doing all this against a background of constant stimulation of the senses and reflex action as well as a complex series of other choices large and small.

Well, if someone can do all that, then it seems that we have a good case for saying that they are exercising choice, acting under free will. Moreover, of course we are usually quite unaware of most of the aspects of the process, since we know only a small part of what is going on in someone's mind. Indeed, the same thing is true of our own choices; part of being 'only human' is acting under motives that we don't consciously think about. So little wonder that we cover our ignorance about the detailed process of choice behind an idea like 'free will'.

That's not to give a full account of free will at all, but rather to point to some things we take to be important about it (including ones that are certainly important to making moral decisions). However, it is to suggest that such things are quite possibly all there is ultimately to free will.

But all those things seem to be something that could be done by a thinking machine—a computer in a broad sense. And of course computers can be built out of physical materials and operated according to deterministic physical laws. That is, it is perfectly reasonable to think that free will is compatible with determinism.

I will leave further discussion of this important topic to the Further Readings. Before we move on, I want to emphasize that, as in chapters 10 and 11, we have seen how viewing ourselves as physical entities in a physical universe allows us to unravel philosophical mysteries: earlier of the passage of time, and now of time travel. Our analysis turned on the fact that we are part of the physical state of the world and subject to its laws, and so the conclusions of the last chapter apply to us as they would any physical system. If there is a portal in the future, then it may constrain the state now, and hence constrain us.

The kind of lessons that physics has for philosophy in this case are somewhat different from those of other chapters, for they concern our perceptions and theories of *ourselves*. A consideration of the physics of perception helped us understand our experience of the passage of time, while a consideration of physical constraints helped us understand our inability to do something we should be capable of. On the other hand, in the previous chapter we asked in general about the compatibility of physics with time travel, without regard to what was time-traveling (hence we ran up against free will here not there). Similar lessons will be learned as we go, and others we will not have time to discuss: Does quantum mechanics shed light on consciousness? If the laws of physics are past and future deterministic, why do we know more about the past?

Further Readings

For a general, popular introduction to the topic of free will, though not in its connections to time travel, I would recommend chapter 3 of Simon Blackburn's *Think: A Compelling Introduction to Philosophy* (Oxford University Press, 1999). Indeed, I recommend this book to anyone with an interest in philosophy. A far more detailed defense of a physicalist view of free will (including a discussion of experimental work) can be found in *Freedom Evolves* by Daniel C. Dennent (Viking Books, 2003).

David Lewis's "The Paradoxes of Time Travel," published in the *American Philosophical Quarterly* (1976: 145–152) is very influential with philosophers but also accessible with some careful thought. The last two sections of this chapter take it as their starting point but attempt to fill in Lewis's response to the paradoxes in a more concrete, physics-oriented manner.

14

Spacetime and the Theory of Relativity

Space, we said, is where everything is located at each moment; spacetime, space plus the dimension of time, is where everything past, present, or future is located. To put it another way, the block universe is spacetime plus all the things that exist or happen in it. Both the block and spacetime make sense without relativity, but since spacetime becomes essential in relativity—there is no unequivocal way of picking out space over time—we shall discuss it in that context.

The purpose of this chapter is to get some understanding of relativity; the core concept for us will be the 'relativity of the present'. It's familiar enough that at different times different things are in the present: dinosaurs during the jurassic period, Plato in antiquity, my breakfast a few hours ago. However, even for two people at exactly the same time and place, what is in the present relative to them may differ; if they have different speeds, then different things will be in their present. From this core concept we shall see why lengths and time intervals are also relative. I want to emphasize that these effects are 'real': changes in length and time are not illusory or merely apparent.

The point of our studying relativity is to see how it changes our concepts of space and time and how it changes the philosophical landscape, especially with respect to the latter. So in the next chapter we will look at two issues in detail. First we will consider why, if someone were to travel away from the Earth and then return, he would have aged less than someone who stayed behind—why he would have 'traveled to his future' in a sense (as was mentioned in chapter 12). Second, we want to see what difference relativity makes to some of the questions raised in chapter 10; how do space and time differ if the idea of the present changes radically?

We've already seen how advances in physics can affect philosophical debates: Newtonian physics forced philosophers to take absolute motion seriously, for instance. This point is well worth emphasizing: scientific developments driven by experiment have philosophical consequences. Relativity makes the point very viscerally because the consequences are still so surprising; strong intuitions about space and time turn out to be faulty philosophical prejudices.

It's also well worth pointing out that Einstein's development of relativity required him to see how notions such as 'simultaneity', 'length', and 'time' were used in physics, rather as Newton and his predecessors had to understand what 'motion' meant in physics. That is, Einstein too gave a philosophical analysis of physical concepts. The new conceptions of these things that we will be studying are the result of that philosophical work.

These two chapters are probably the most demanding, and we will need to learn a new kind of geometry to explore the consequences of relativity. I suggest having some paper and a pencil and ruler handy, to draw your own figures as we go. One more thing: we will discuss first the 'special theory of relativity' rather than general relativity. One way to understand the difference is that special relativity assumes that spacetime is not curved and that the distribution of matter makes no difference to that fact. I will say more about the differences later.

14.1 PHOTONS AND BULLETS

Our discussion will turn on an astonishing, mind-bending fact about light: it travels at 3×10^8 m/s relative to everyone however they are moving, and however the source of the light is moving. (That's the speed in a vacuum; it's close enough to the speed of light in a gas such as our atmosphere.)

To see how astonishing this fact really is, compare light from various sources—light bulbs, sparks, the sun, and so on—to bullets from a gun. To make this comparison as close and clear as possible, we will focus on a single particle of light, a 'photon'; when light shines from a source, a stream of photons is emitted, much as a gun emits a stream of bullets. (According to quantum mechanics, photons also behave like light waves, but it's easier to think of moving particles.)

First, the speed of a bullet depends on the gun firing it; different guns have different muzzle velocities (the relative speed at which a bullet leaves the gun), and of course a bullet fired from a moving gun travels faster (relative to you) than one fired from a stationary gun. Not so for light; everything that emits light emits photons at the same speed, regardless of whether it is a light bulb, a spark, the sun, or a quasar, and however fast it is moving.

That might not seem so remarkable. Something similar is true of sound, for instance: whatever the motion (or nature) of their sources, all sound waves travels at the same speed through any given medium—air or helium or water or whatever (what speed depends on the medium). However, a source-independent speed is an extraordinary property of light, for it holds in a vacuum, where there is no material medium! So light's source-independent speed can't be a matter of a speed fixed relative to any material thing (such as an 'aether', as people thought at one time). Instead it must be the structure of spacetime itself that determines the motion, much as we saw spacetime structure play a role in Newtonian mechanics in chapter 9.

But that is not the most astonishing thing! The most astonishing thing is that whatever the source and however fast it is moving *and however fast you are moving*, light moves at 3×10^8 m/s relative to you. This behavior is of course entirely unlike that of bullets. If you run after a bullet, its speed relative to you is less than if you run toward it. It is also unlike that of sound: for instance, sound travels at 0 m/s relative to a plane flying at mach 1. But if you run after (or toward) a photon, it still moves at 3×10^8 m/s. (And so, if you thought after the last paragraph that the structure of spacetime which determined the motion of light was a standard of *rest*, relative to which all light travels the same speed, you'd be wrong. Because then you would see photons travel at 3×10^8 m/s only if you were at rest, otherwise slower or faster, depending on how you moved.)

Concretely, suppose that a certain gun fires bullets at say, 100 m/s; that is, the bullets travel at 100 m/s *relative to the gun* and the shooter. Then, relative to a bunny rabbit hopping away from the gun at 10 m/s a bullet travels at 100 m/s − 10 m/s = 90 m/s, while relative to another bullet also fired at 100 m/s *at* the shooter by a duelist, the speed of a bullet is 100 m/s + 100 m/s = 200 m/s. (One consequence of relativity is that simply adding relative speeds in this way is not 100 percent accurate, but the basic point holds: the relative speed of the bullet depends on the relative speed of the gun.) Thus the speed of the bullet is different relative to the shooter, bunny, and second bullet. Because they have different speeds relative to one another and relative to the gun, the bullet has different speeds relative to each.

But light does *not* act like that. First, if you measure the speed of light from sources moving at different speeds you will find that it is always the same, always 3×10^8 m/s, *not* 3×10^8 m/s plus the speed of the source relative to you. Idealizing, just time how long photons from various sources—a stationary light bulb, a headlight on a moving rocket, the sun, and distant stars—take to travel down a 3×10^8 m track. They all take 1 s.

Again, the same result (at a slower speed) also holds for sound, because the speed of sound is fixed by the medium in which it travels. But light is not like sound either, because one and the *same* photon has the very same speed relative to anyone, regardless of her motion. If you are moving relative to me and we both measure the speed of the *same* photon, I will find that it is moving at 3×10^8 m/s relative to me, and you will find that it is moving at 3×10^8 m/s relative to you. Idealizing again, suppose we both are carrying 1 m rulers (so your ruler moves relative to mine), and we both time how long the photon takes to travel down our own ruler; we will both clock it at $1/(3 \times 10^8)$ s. (Real experiments are a bit more complicated but in essence do the same thing.)

Again, that result is of course completely unlike that of bullets or sound. A bullet speeding away from me does not move away from Superman, who can even catch up with it, being even faster. Similarly, a supersonic aircraft can overtake the sounds I make. The bullet has a different speed relative to me and to Superman, and the sound relative to me and to the plane. But however fast Superman or the plane chases after a photon, it moves away from them and me at 3×10^8 m/s.

These facts are the crux of our treatment of relativity. Indeed, the fact that the speed of light from any source is the same relative to everybody captures a lot of what is *relativistic* about relativity: it means that our relative motions make no difference to the speed. So consider another example to drive the point home. Imagine a train rushing past you in the dark, with its headlights on. The light from the headlights travels at 3×10^8 m/s relative both to the train and to you, however fast the train moves past—at 10 m/s, 100 m/s, 1 km/s, 10^8 m/s . . . ; a bullet traveling at 100 m/s ahead of the train would travel at 110 m/s, 200 m/s, 1.1 km/s, . . . (again, that's not 100 percent right).

That light could be like this is so counterintuitive, so contrary to experience, that when people learn relativity they often assume that whoever is teaching them isn't really serious, or is speaking in some odd way (some special sense of 'speed' not like the normal one), or perhaps that they are just plain crazy. And so people who have been taught relativity often don't come to grips with what relativity is really saying. So let me emphasize for the record that I really mean to say that in the perfectly ordinary senses of the words, absolutely any light, from whatever source, however it moves, always has exactly the same speed relative to everybody, however they move.

If that idea has sunk in, then you may want to deny that what I am saying is actually true; like death, relativity has stages of acceptance. In that case I refer you most directly to the real experiments that measure the speed of light, and to the fact that vast swathes of modern physics simply would not be correct if relativity were not—and relativity tells us that all light travels at 3×10^8 m/s relative to anyone. The 'principle of the constancy of the speed of light' is experimental fact.

The constancy of the speed of light may be a hard principle to swallow and may even appear logically impossible. But if we shift our understanding of space and time, in a way that is quite counterintuitive but perfectly logically consistent, then the principle of the constancy of the speed of light is coherent.

For instance, imagine an astronaut entering a darkened room on a spaceship bound for Mars and shining a torch (you might call it a 'flashlight') on the far wall, 30 m away. A photon from the torch travels at 3×10^8 m/s relative to her, and so it reaches the far wall in $30/3 \times 10^8 = 0.0000001$ s relative to her. Relative to the Earth, however, the photon has to travel 10 m *plus* the distance traveled by the rocket in the meantime, before it reaches the far wall; the photon is 'chasing after' the wall (nothing peculiarly relativistic about that). Since, by the the principle of the constancy of the speed of light, the photon (the same photon that we were considering before) also travels at 3×10^8 m/s relative to the Earth, but over a longer distance, it must take a different amount of time—*longer* than 0.0000001 s!

That is, we can start to make sense of the principle if we accept the counterintuitive idea that how long a process takes depends on one's speed; one and the same photon takes less time to reach the far wall relative to the rocket than relative to the Earth. The goal of the following sections and the next chapter is to understand in more detail how our conceptions of time and space must change in order to accommodate the principle of the constancy of the speed of light. First, a confession.

14.2 CONVENTION

I swept a philosophical controversy under the carpet in what I just said, and it would be misleading if I didn't acknowledge its existence. That said, this section isn't crucial for understanding what follows, so if you would rather skip it, it won't affect your understanding of the rest of the chapter.

Suppose we ask, as Poincaré did about space, what the basic experimental facts are. Without going into details, in both cases the answer is that phenomena pick out a *group*. For space, the group was that of the spatial displacements of rigid bodies. For relativity, we are concerned with time as well as space, so the group involves both and is more complex: first, it involves displacements in space *and* time, which means changes in speed; second, it involves things extended in time as well as space, which means systems while they evolve. So the group established by the basic experimental facts is of the spatiotemporal displacements of spatiotemporal systems.

The mathematics is more complicated, but the idea remains the same: the group corresponds to the displacements of a geometry, now of time and space, or rather spacetime. This is the geometry that we will explore here. In every application of relativity in physics it is taken to be the geometry of spacetime. According to Poincaré's argument, though, physicists do not thereby draw a conclusion from experiment but choose a convention from many possibilities. Part of what they choose is that light travels at the same speed in *every direction*; according to other choices its constant speed would depend on the direction it traveled. Clearly the physicists' choice simplifies things, and it is the one that we will adopt.

So, according to Poincaré's views, the experiments I described, the principle of the constancy of the speed of light, and the things that follow from it, involve a convention (and of all the things that follow, the rule for relativizing simultaneity is the one most discussed by philosophers). As we saw, Poincaré's views are controversial, and so there is a philosophical controversy about their applicability to relativity. But even if he is correct, what follows? Is everything we say just a matter of defining words in a new way? No.

First, the principle obviously involves conventions other than a choice of geometry, because the speed of light is given in m/s not, say, miles/hour (it's about 670,616,629 miles/hour). A dispute over which units are correct is not a dispute over the speed of light, but at most over which units are better. That choice doesn't diminish our ability to make a substantive statement about the speed of light. Indeed, it is the convention that allows us to make it!

The same goes for the geometric convention. One cannot 'disprove' other choices, but a choice has to be made in order for one to make substantive claims about space and time in relativity. If another choice were made instead, then we would end up describing the same facts but in different (and frankly more awkward) terms. The fact that we have to choose a geometry does not make everything that we say a matter of definition; instead it is what allows us to make substantive claims at all about space and time in relativity.

So what is most important to understand is that our astonishing results about space and time are not the result of choosing a new and bizarre convention for describing our familiar, everyday expectations about the behavior of light, bodies, and clocks. Instead, experiment shows that things behave quite differently (for instance light behaves neither like bullets nor like waves in a medium). As before, new discoveries lead to new conventions. What we explore below are the new behaviors, and we use the new conventions to do so.

Let me say one more time, the behaviors are *new* (and bizarre!) and are not explained away by the choice of convention. There is no way of getting rid of relativistic effects by making a change in convention; ultimately, experiments just come out differently in relativity than from Newtonian physics. The same point holds for the geometry of space: different choices give different meanings to 'straight', but they don't change whether measurements of a given figure satisfy the Pythagorean theorem.

14.3 RELATIVITY—WHEN IS NOW?

We will start with the 'relativity of simultaneity': roughly, people in relative motion will disagree about which events occurred simultaneously. Was the astronaut's torch turned on at exactly 1 P.M. in Chicago? Or before? Or after? It depends on the speed of the person whom we ask.

However, there is something potentially misleading about putting things this way; it suggests that *people* are somehow especially relevant to the story, when they are not. By referring to people, this gloss suggests that the 'disagreements' are a matter of 'opinion', like disagreements about which is the best ice cream flavor, or superhero, or James Bond. These depend on your 'point of view', but not in the way simultaneity does. Such disagreements about ice cream and so on depend on your subjective psychological state, but nothing like that is going on in relativity; which events are simultaneous is not a matter of subjective 'opinion' but a fact about the physical world that depends on one's state of motion.

So to avoid confusion on this point, we will not talk about 'people' (or 'observers') but will discuss simultaneity relative to 'reference bodies'. As in chapter 9, all material objects (including human bodies) count as potential reference bodies. They serve as possible standards for specifying positions and motions (and simultaneity), as positions and motions (and simultaneity) relative to the reference body in question. Then the relativity of simultaneity can be glossed by saying that whether two events occur simultaneously is relative to this or that reference body and depends (in part) on its speed. Thus the 'disagreements' don't depend on how things are perceived, or subjective psychology, or people at all, any more than 'disagreements' about the relative speed of bullets do; what matters is the motion of the body to which simultaneity or speed are referred.

With the idea of a reference body in hand, let us see why simultaneity must be relative, and indeed what this really means. Consider another thought experiment, after one given by Einstein in his popular expositions of relativity. Imagine a light bulb sitting in the middle of a train car, and light detectors at either end; when the bulb is switched on, light (i.e., photons) travel from it, some in the direction of the front of the train and some toward the rear. Since the bulb is at the center of the car, the distance to either target is the same, so, since light travels at the same speed in both directions—at 3×10^8 m/s—the light will reach the detectors *simultaneously*. At least that is the story relative to the car, our first reference body. (See figure 14.1.)

Suppose that while this experiment takes place, the train is passing through a station, our second reference body. Relative to the station (and to the stationmaster standing on the platform) the detector at the rear of the car is moving toward the photons shone backward, while the detector at the front is 'running away from' the photons headed forward. Thus relative to the station, the distance that the light has to travel to reach the target at the rear is less than the distance that it has to travel to reach the target at the front. There's nothing relativistic about that, of course. But because of the constancy of its speed, the light travels at 3×10^8 m/s *in both directions* relative to the station. Therefore, relative to the station, since the forward traveling light has further to go than the rear traveling light, yet they move at the same speed, the light must reach the front target *after* the light reaches the rear.

That is, relative to the train, the light reaches the ends of the car at the same time, while relative to the station, it doesn't. All of the reasoning to this startling conclusion is perfectly normal and intuitive, except for the premise that light travels at the same speed relative to all reference bodies. Intuitively—imagining photons to be like bullets, say—we expect that, relative to the station, the forward traveling light will move at the speed of light *plus* the speed of the train, and rear traveling light at the

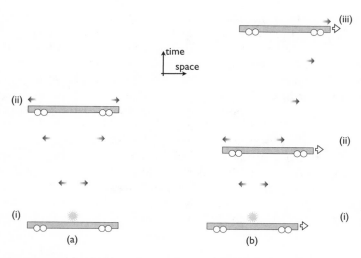

Figure 14.1 (a) Relative to the train (i) two photons are emitted simultaneously from a bulb, at the center of the car, so (ii) since they move at the same speed, they arrive simultaneously. (b) Relative to the station (i) the photons (the very same photons) are emitted simultaneously, at the center of the car, but since the train is moving (ii), the rear-moving photon has to travel less distance than (iii) the forward moving photon. Since they travel at the same speed, the rear photon arrives first.

speed of light *minus* the speed of the train. So we intuitively expect the photons to be speeded up and slowed down by just the right amounts to arrive simultaneously relative to the station too.

But the constancy of the speed of light means that that is not what happens; instead they travel at the same speed and so arrive at different times. Reflect on how startling that conclusion is. Intuitively, if two events occur simultaneously, then that doesn't mean 'for this or that person' or 'this or that reference body', it means for everyone and everything. The relativity of simultaneity implies that this intuition is just wrong: events that occur at the same time relative to the train and its passengers do not occur at the same time relative to the station and the stationmaster. For instance, if two clocks tick in time relative to the train, then they are not synchronized relative to the station; or if the brakes are all applied at once relative to the train, then relative to the platform they are applied in sequence from rear to front.

Let me stress that in these examples it is the same events—a pair of photons arriving front and back, clocks striking one, brakes being applied, and so on—that occur simultaneously relative to the train but not simultaneously relative to the station. I am *not* talking about merely similar events occurring on separate occasions.

Let me also stress that I *don't* mean that the stationmaster merely *sees* the light striking the rear before the front. Indeed, if he is standing so that the front of the car passes him as the light strikes it, then light reflected from there will reach him before light reflected from the rear; he will *see* the front illuminated before the rear, even though relative to him light reaches the rear first. (There's nothing odd about that: imagine fighting a three-way duel with opponents at 50 m and 100 m, with the latter firing just before the former, so that the light, sound, and bullet from the nearer arrive before those of the further. The first shot that you see, hear, and receive is the one that was fired second, just because light, sound, bullets and indeed everything take time to reach you. When something arrives depends on when and *how far away* it was emitted; there's nothing peculiarly relativistic about that.) Relativity says that the order in which events occur is relative, not just the order in which we find out about them.

14.4 RELATIVISTIC SPACETIME

If two events are simultaneous, then they are (at most) separated in space, and if they are nonsimultaneous, then they are separated in time. Thus the relativity of simultaneity means that the very distinction between space and time is relative in the following sense: whether the 'gap' between events is just spatial or also temporal is relative. We shall explore two surprising consequences for the nature of space and time: length and the passage of time are relative.

Relativity is in part a theory that proposes a new geometry, involving time as well as space; the consequences of relativity can be explored using geometric figures, much like those familiar from Euclidean geometry. These figures give a very useful and clear picture of spacetime and simultaneity relative to different reference bodies; they show how spacetime is divided up into space and time differently by different reference bodies. Picture, then, spacetime with time running up the page and space across. Pick a reference body: you, for instance, or the Earth, or the center of our galaxy. Draw its worldline straight up the center of the figure, and take any points on a horizontal line to represent events that occur simultaneously relative to it. See figure 14.2. There's nothing relativistic about this diagram: we would intuitively picture spacetime in just this way.

Next we want to picture other things moving around in spacetime. It is convenient to adopt the convention that the paths of photons will be at 45° to the horizontal. Since light travels very fast by familiar standards, it follows that the scale is not a very familiar one. If, for instance, we choose a scale in which each centimeter across the page represents 1 m, then each centimeter up the page must represent the time it takes light to travel 1 m, namely $1/3 \times 10^8$ s! Or if each centimeter up the page represents 1 s,

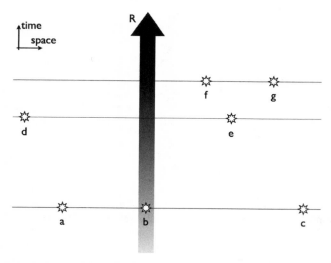

Figure 14.2 The worldline of the reference body, R (often omitted for clarity in future diagrams), is the vertical arrow. The horizontal lines represent events that occur simultaneously relative to the reference body; e.g., whatever happens at a, b and c happens simultaneously relative to R, similarly for d and e, and for f and g (but not for a, d and f).

then we have a scale in which each centimeter across the page represents the distance traveled by light in 1 s, namely 3×10^8 m. When we consider lengths, it will be convenient to use a scale like the former one, so a picture of a 1 m body is big enough to see. When we consider times, the latter will be better so that we can picture a 1 s interval. Either way, a body traveling slower than light will have a worldline that is drawn closer to the vertical than 45°, since it travels less distance in space than light in any given time. An example of a spacetime diagram is given in figure 14.3.

Finally, which points of our picture represent events that occur at the same time relative to a moving body? Not ones that are on a horizontal line, for they are simultaneous relative to the original reference body. There's a precise rule that answers the question: *two points represent events that are simultaneous relative to a body if they lie on the reflection of its worldline in a line running at 45°*. Equivalently, the trajectory of a photon bisects the worldline of a body and the line of events simultaneous (relative to the body) with the arrival of the photon at the body.

It's not hard to show that this is the correct rule, but unnecessary for us (see Further Readings). You can see from figure 14.4 that the rule implies horizontal 'lines of simultaneity' relative to the stationary reference body, and the simultaneous arrival of photons back and front relative to the train car, as it must.

(One remark: which worldline runs vertically is a matter of free choice in drawing these figures and does not signify any special physical property.

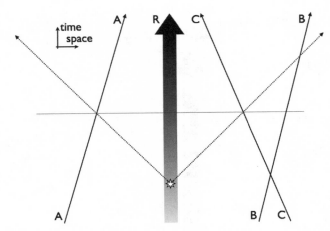

Figure 14.3 A, B, and C represent the worldlines of three objects. They cover less distance in space in any given time than the (dotted) worldlines of the two photons (at 45⁰); A, B, and C travel slower than light and so have steeper worldlines than photons. A flash of light at the point of space and time shown emits photon left and right. C, moving to the left relative to R, experiences the following events: first he passes B; then he passes a photon coming from the left, at the very same moment relative to R that the other photon passes A. C will soon meet A, but he will never reach the second photon, since it is moving faster to the left than he is. And so on.

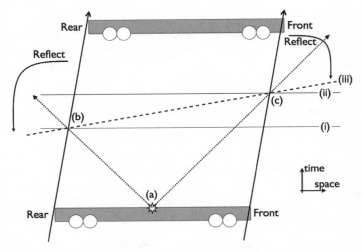

Figure 14.4 In this diagram we see some (horizontal) lines of simultaneity ((i) and (ii)) of a stationary reference body—the railway station. We also see a train car moving from left to right, and two photons being emitted at the center of the car (a) and arriving at the rear (b) and front (c). Since (i) is before (ii), (b) occurs before (c) relative to the station; the rear of the train catches up with the left-traveling photon, so it reaches its target first. However, relative to the train, (b) and (c) occur simultaneously, so a line of simultaneity relative to the train (iii) passes through them. Simple geometry means that the photon worldlines bisect the train worldline and (iii).

Just as every Euclidean figure could be redrawn at a different angle without changing any geometric properties, so every one of our pictures could be redrawn with a different vertical worldline without changing any of the physics pictured, provided photons still have 45° worldlines and the same rule for lines of simultaneity is used.)

There's nothing very tricky or complex about these 'Minkowski diagrams' (after Hermann Minkowski, the first to explain relativity geometrically, and from whom I took the title of this book). The only unfamiliar step is drawing in the lines of relative simultaneity carefully. But they are sufficient to demonstrate a number of startling features of relativity. First of all, bodies shrink as they speed up.

14.5 RELATIVITY OF LENGTH

Imagine driving along and deciding to find out whether a car headed toward you was longer, the same length, or shorter than yours. One way to do so is to wait until the front of the oncoming car is level with the rear of your car and then determine whether at that moment its rear sticks out beyond, is level with, or is within the front of your car.

Let's draw a Minkowski diagram for a situation like this. Suppose we have two identical cars, A and B. In fact, suppose that before the experiment they were parked next to each, with their fronts and rears aligned, so they were the same lengths (5 m perhaps). Imagine next that they are driven away in opposite directions, and then head back at equal speeds (without any other changes to them). Equal speeds means the slopes of the worldlines are the same (equal distances in equal times), but in opposite directions.

We have to be very careful about our assumptions about lengths and times; the constancy of the speed of light means that our intuitions are systematically wrong. So how long should we picture the cars to be? Well, let's suppose that there is no special direction to space, so that the lengths depend only on the speed that the cars are traveling and not on the direction. Thus relative to the reference body, A and B are the same length, whatever that is. Then all we need to do is draw the cars the same length: see figure 14.5

Look at the point in spacetime at which the rear of A is aligned with the front of B. Relative to the reference body of the picture, events simultaneous with this point lie on a horizontal line, as the point at which the front of A is aligned with the rear of B does. Since A and B line up relative to the reference body, they are indeed pictured the same length relative to the reference body.

Now draw in the line of events simultaneous with the same point, this time relative to A: the reflection of the worldline of A in a 45° line through the point. We see that relative to A, when the rear of A is aligned

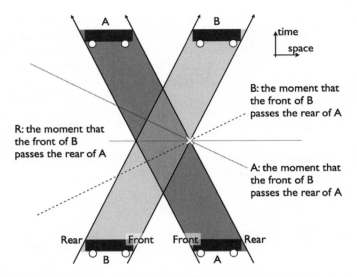

Figure 14.5 Identical cars A and B compare each other's lengths as they pass at equal speeds (relative to R, which is not shown) in opposite directions. X marks the point in spacetime at which the front of B passes the rear of A. Along the line of events simultaneous with X relative to A, B is within the length of A; along the line of events simultaneous with X relative to B, A is within the length of B. Each is shrunk relative to the other! Relative to R they are equally long: the front of B passes the rear of A at the same time relative to R that the rear of B passes the front of A.

with the front of B, the rear of B is *within* the front of A; relative to A, B is shorter than A!

And that's not all. Now draw in the line simultaneity relative to B. We see that relative to B, when the rear of A is aligned with the front of B, the rear of B *sticks out beyond* the front of A—so relative to B, B is longer than A!

But when these cars are at rest next to each other they align perfectly: they are the same length. So the only conclusion is that the cars—and hence any objects, since there's nothing special about cars—shrink as they get faster. That is, in addition to distance, speed, and simultaneity being relative to a reference body, so is length! If you are asked 'how long is that car?' there is no absolute answer: you have to ask first, 'relative to what?' before you can reply.

How deep and shocking is that conclusion? Looked at one way, it seems not that much. Consider, for instance, A's actions relative to a stationary reference body, one with horizontal lines of simultaneity.

Relative to a stationary body, A compares the place where his rear is aligned with the front of B, to the place of the rear of B a little time later *after B has moved to the right*, and to the place of the front of A a little time even later *after A has moved further to the left*. Relative to the reference body, A has considered not the length of A but the distance between the rear of A at one time and the front of A a bit later, which of course is longer than A, since A is moving forward!

And relative to the reference body, A has considered not the length of B but the distance between the front of B and the rear of B a bit later, which is of course shorter than B, since B is moving forward. So relative to the reference body, the distances that A compares are not the lengths of A and B relative to the standard of rest of the picture, but lengths a bit longer and a bit shorter than A and B respectively. Of course these lengths are different even though the lengths of A and B are equal relative to the standard of rest!

But that is not to say that A's measurements are wrong, just wrong relative to the standard rest of the picture (or relative to B). Relative to A they are exactly right; if he wants to know the length of B relative to him, he has to consider the positions of the front and rear of B at the same time relative to him. And so we see very clearly that the disagreements about lengths between A, B, and the standard of rest are due to their disagreements about which events are simultaneous—that the relativity of length follows from the relativity of simultaneity.

Hence length contraction is as deep and shocking as the relativity of simultaneity, which, as we made clear, is deep and shocking. The relativity of length too is not a matter of subjective psychology, or 'opinion', but as real as space and time themselves.

Let me emphasize the point with the following important fact. When we discussed the constancy of the speed of light, we considered the simple experiment of timing a photon down a 1 m ruler. What I didn't stress then is that the same result will be obtained *however* the speed is measured. Use any kind of ruler that is rigid according to the laws of physics, and any kind of clock that keeps good time according to laws of physics, and you will find 3×10^8 m/s (regardless of source or who measures it). That is, the laws of physics imply the constancy of the speed of light.

But the relativity of length follows, via the relativity of simultaneity, from the principle of the constancy of the speed of light, and so follows from the laws. And this is exactly what one finds if one considers the laws in detail. They dictate, for instance, that the intermolecular electric and magnetic forces holding a moving body together will cause it to shrink compared with an identical body at rest. And that is as objective and real as one could possibly get.

We need, however, to be careful of course, because the effect is reciprocal. The point is that the laws entail different lengths for bodies relative to different reference bodies: relative to A, B is moving and hence shrunk by its intermolecular forces, and vice versa. That is, those forces are themselves relative to a reference body.

14.6 RELATIVITY OF TIME

Now for the relativity of the passage of time. Imagine two identical clocks, C and D, moving in opposite directions at the same speed; suppose that they both read exactly noon as they pass. Since they are identical, when they are both stationary then they run at the same rates—one tick per second, say. The question is how much time passes on one in the time it takes a second to pass on the other: when C reads 12:00:01, what does D read?

First, at what point do C and D read 12:00:01? We can't trust our intuitions about lengths and times, but because the clocks move at the same speeds, and because the direction of motion is irrelevant, the same length along each worldline represents the passage of the same amount of time on each clock. So if we arbitrarily pick a point on C's worldline as 12:00:01 on C, a point the same distance along D's worldline represents where D reads 12:00:01. All this is shown in figure 14.6.

To find out what D reads relative to C when C reads 1 s past noon, we need to consider what point along D's worldline is simultaneous relative to C, with the time that C reads 12:00:01. From the diagram, clearly it is *before* D reads 12:00:01. And vice versa. So each clock runs slow relative to the other; again it's clear that the disagreement arises because of the relativity of simultaneity.

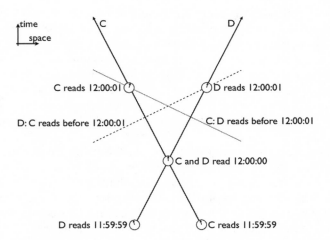

Figure 14.6 Two identical clocks, C and D pass at equal and opposite speeds as each reads 12 o'clock. Since their speeds are equal, each reads one second later at equal distances further along their worldlines. Lines of simultaneity relative to each clock, when each reads 12:00:01, are drawn. The point on D's worldline simultaneous, relative to C, to the point at which C reads 12:00:01 is before the point at which D reads 12:00:01. Thus D runs slow relative to C. And vice versa.

It's important to appreciate that the slowing down of moving clocks is not due to a defect in their running but because of the relativity of time itself. The clocks are working perfectly according to the laws, but they read differently because how much time elapses between two points is a relative matter. As before, the effect follows, via the relativity of simultaneity, from the constancy of the speed of light, which follows from the laws. According to the laws, whatever forces hold the parts of a clock together and cause it to 'tick' at a fixed rate will be changed when the clock is moving, in such a way to make it slow down. (Of course the effect is again reciprocal: the laws make C run slow relative to D and vice versa.)

Although D runs slow relative to C, we can ask another question: does someone traveling with C and watching D *see* it running fast or slow? Well, we see events when light arrives from them: you don't see the bulb turn on, or the firework explode, or a sunspot appear until light from those events reach your eyes, after the events occur. (In the first two cases your brain takes much longer to become conscious of the light arriving than it takes for the light to reach you.) So what we want to know is how much time elapses between the moment light reaches C from one tick of D and the next.

Consider then the paths of photons that leave D at 11:59:59, 12:00:00, and 12:00:01, and how much time elapses between their arrivals at C—the times at which C sees the corresponding ticks. They are shown in figure 14.7.

Again the clocks differ only in the direction of motion, so *equal distances along their worldlines mean equal time intervals on the clock*. Then we can see from the diagram that as measured on C, the time between the moment C sees D showing 11:59:59 (the point the third photon

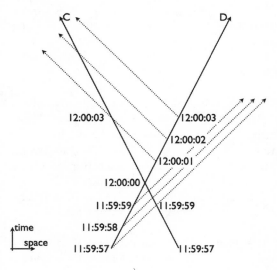

Figure 14.7 Identical clocks C and D again pass at equal and opposite speeds. The worldlines of photons leaving D show when someone traveling with C would see D read the times indicated: before C and D meet, less than 1 s passes on C between the arrival of each photon, so C sees D run fast; afterward, more than 1 s passes on C between the arrival of each photon, so C sees D run slow. As we just saw, in relativity, at every stage D is in fact running slow relative to C.

arrives) and the moment C sees D showing 12:00:00 (the point they pass) is less than 1 s. (Indeed, the picture shows that C sees 11:59:57— the first photon—at 11:59:59.) Therefore, as D moves toward C, C sees D running fast, even though D is running slow relative to C! But this situation is not very surprising: when D is moving toward C, light has less distance to travel from successive ticks, so of course it arrives more quickly, and so the ticks are seen closer together than they occur.

Similarly, the time on C between the clocks passing (noon) and when C sees D read 12:00:01 is more than 1 s; the second photon arrives after C says 12:00:01. Thus not only does D run slow relative to C, but C sees D run slow as it moves away. (However, C sees D running even slower than it actually runs relative to C: relative to C, at the moment C sees D read 12:00:01, D actually reads a later time!)

This phenomenon—the 'Doppler effect'—does not depend on relativity or the constancy of the speed of light. We would see clocks moving toward us running fast and clocks moving away running slow even if the speed of light were not constant; all that matters is that light travels at a finite speed over decreasing or increasing distances. (What would be different would be the rate at which such clocks would appear speeded up or slowed down.) And of course the phenomenon is symmetrical: D sees C running fast as they approach, and slow as they part.

The Doppler effect means it would be misleading to say that relativity implies that we *see* moving clocks running slow. Taken literally, this claim is false: we see clocks moving toward us run fast. And even in situations in which it is true—for clocks moving away—it would be true even if relativity were not. (Physicists do sometimes say 'A sees B shrunk' or 'C sees D running slow', but they do not necessarily mean 'sees' in the sense of an optical experience. They use it as a shorthand for 'B is shrunk relative to A', and so on. To avoid confusion, I have avoided such loose talk.)

Further Readings

My favorite formal—but not very—introduction to special relativity is David Mermin's *Space and Time in Special Relativity* (Waveland Press, 1989). It requires more algebra and trigonometry than I have used, but it is very clear about the interconnections between the ideas, and it fills in some of the gaps in our discussion. He has also recently produced a sequel, *It's About Time* (Princeton University Press, 2005).

Einstein's proof for the relativity of simultaneity is found in chapter 9 of his *Relativity: The Special and General Theories* (first published in 1920 by Henry Holt).

A nice introduction to the topic of conventionalism in relativity is found in Wesley Salmon's *Space, Time, and Motion: A Philosophical*

Introduction (University of Minnesota Press, 1980). I'd also like to note that the geometry to which I take the spacetime group to correspond is that of Minkowski spacetime, plus the standard definition of simultaneity (which is uniquely definable in terms of Minkowski geometry and a timelike vector).

15

Time in Relativity

In the previous chapter we learned the basics of relativity through a simple, but non-Euclidean, geometry. It should be clear that the theory plays havoc with our expectations of space and time, but in this chapter we will investigate in more detail a couple of consequences, first for 'time travel', in a sense, and then for the conception of the present. I want to make the point that advances in physics really can have profound philosophical implications. (It's also true that the development of relativity required deep philosophical analysis from Einstein; that story can be explored in Further Readings.)

15.1 THE TWINS

Consider the circumstances mentioned at the start of chapter 12, in which my son Kai travels away from the Earth, then returns, taking 10 years, finding that his twin brother Ivor has aged 20 years. For a suitable trip, this result is exactly what relativity predicts. Let's see why.

As Kai moves away, we know that relative to Ivor his time passes slowly, and so at any moment relative to Ivor during the outward trip, Kai will have aged less than Ivor has. We are clocks of a certain kind, so like other clocks, we are affected by the relativity of time. As Kai moves back, he is still moving relative to Ivor, so again relative to Ivor, time passes more slowly for Kai and he ages more slowly. Since relative to Ivor, Kai ages less than Ivor does during both legs of the trip, he will have aged less than Ivor during the whole trip. He must return older than when he left and younger than Ivor is when they are reunited, but other than that he can come back any age, if he takes the appropriate trip.

And there Kai is, standing (not in relative motion now) next to Ivor, obviously 10 years younger, say (he'd have to travel at nearly 90 percent of the speed of light—a round trip to the star Sirius—to pull this trick off). The effect is absolutely real, not merely a matter of 'seeing' clocks run slow. (What does Ivor see if he watches Kai? Because of the Doppler effect, each sees the other aging slowly as they move apart, and fast as they move together. Consistent with the disparity in their ages at the end,

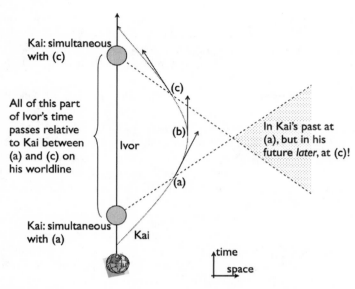

Figure 15.1 Ivor's twin, Kai, takes a trip from the Earth and back; when they are reunited, Kai is younger. To find Kai's lines of simultaneity we must use the tangents to his worldline; those at (a), (b) and (c) are shown, together with appropriate lines of relative simultaneity (it's horizontal at [b]). Because Kai's worldline is not straight, we see a region in which events go from his relative past to relative future! We also see (qualitatively) that considerable time passes for Ivor relative to Kai during the rather short-looking part of Kai's trip between points (a) and (c).

Ivor sees Kai aging slowly for more than half the time. You could check this by adding photon lines to figure 15.1.)

Indeed, this phenomenon is not restricted to interstellar travel but is happening all the time, though generally to an unnoticeable degree. For instance, if someone runs 100 m away from you and then runs the 100 m back, taking 30 s, he will also have aged less than you by 37 quintillionths (US) of a second. That difference is far too little to notice but real, and that is quite a lot to swallow. (Note that the effect is detectable in laboratory experiments.)

Since this phenomenon cannot be put down to mere appearance, it is often a point at which people rebel against relativity; such a possibility seems wildly counterintuitive. There is plenty of reason to believe that intuition, not relativity, is wrong, but the example seems to involve a contradiction, which would be fatal to the theory.

That is, the relativity of time is symmetrical, so Ivor also ages slowly relative to Kai, and the same argument apparently holds: relative to Kai, Ivor ages less during both legs of the trip and so should be younger when they are reunited. But they can't both be younger than each other! What has gone wrong, relativity or some step in this argument? The answer is the latter; there is an important detail, implicit in the story, that breaks the symmetry between Kai and Ivor, so that the only relativistically valid conclusion is that Kai is the younger when he returns. Let us work through the situation more carefully, using what we have learned of Minkowski geometry.

The crucial difference is that Ivor's worldline involves no acceleration, while Kai's trip requires him to come to a stop and then accelerate back toward the Earth in the middle. We saw, in chapter 9 that Newtonian mechanics draws a real, physical difference between different rates of acceleration, which distinguishes rotating from nonrotating buckets, for instance. The same distinction holds in relativity; we will represent it in our geometrical account of the theory by stipulating that *the reference body of a Minkowski diagram must not be accelerating*.

In the present case, that means that figure 15.1 is an allowed Minkowski diagram of our story, but a diagram that had Kai vertical would not be, because he does accelerate. (Ivor is on the Earth, which orbits the sun, so he is in fact accelerating, but we can treat him as approximately nonaccelerating.) Moreover, since the reasoning concerning their ages depends on conclusions about clocks drawn from Minkowski diagrams in the previous chapter, it only applies relative to Ivor.

But to show that there is no contradiction, we should work out what Minkowski geometry predicts about Ivor relative to Kai. We can't draw a Minkowski diagram with Kai as the reference body, since he is accelerating, but we can still draw in lines of simultaneity for him in figure 15.1. If you want to know what events are simultaneous relative to an accelerating body, you have to consider the trajectory it would follow if it suddenly stopped accelerating. Geometrically, that means looking at the tangent to its worldline; the simultaneous points are those along the reflection of the tangent in the worldline of a suitable photon.

These have some very counterintuitive properties. Consider the lines of simultaneity relative to Kai, on the way out just up to the time he changes direction, and on the way back starting just after he starts to return (points [a] and [c] in figure 15.1). They are not parallel; some events that were in the past relative to Kai at (a) are in the future at (c), and of course simultaneous with Kai before and after these points! It's enough to make you wonder whether relative simultaneity really amounts to a 'present' after all.

Similarly, events along Ivor's worldline that were well in the future of Kai have become in the past relative to Kai rather rapidly, while he was accelerating in fact. As Kai accelerates, relative to Kai, Ivor's time passes very fast. So qualitatively, the diagram shows Ivor aging more than Kai during the trip—removing the threat of contradiction. (In the picture, Ivor and Kai's worldlines represent 20 and 10 years respectively, even though Kai's worldline is longer than Ivor's; clearly equal distances along these worldlines do not represent equal times. Even so, the impression is correct; a quantitative calculation of the length of Ivor's worldline relative to Kai shows that during the time that Kai accelerates, relative to him a great deal of time passes for Ivor.)

So certainly, relativity as captured in Minkowski geometry is not shown to be inconsistent by the twins; it's only a 'paradox' in the sense of being surprising, not contradictory. But it isn't clear that the foregoing fully *explains* what is happening. Well, as we have discussed, explanations can ask for different things, but one response is to say that it is simply the laws of physics governing Kai and Ivor (and everything else) that explain what goes on. They describe the forces and motions of all physical things and so govern how 'clocks'—even biological ones, like people—tick. So, concretely, one can work out, say, relative to Ivor, how both he and Kai age, and one will find that what they predict is a 10-year age gap. (Or if you prefer, suppose they carry simple clocks with them; the laws predict that relative to Ivor, say, the various forces governing the clocks will be such that Kai's clock will advance 10 years while Ivor's advances 20.) The explanation is that that is what the relativistic laws of physics predict for the behavior of clocks.

By the way, what does Kai see if he watches Ivor? He sees Ivor aging slowly for the outward half of his trip, and fast for the return half; of course, 20 years in total.

15.2 GENERAL RELATIVITY

So far we have restricted attention to special relativity. But the general theory of relativity is our best-developed physical understanding of space and time, so we must ask what it has to say. The special theory is (in a sense) a much simpler approximation to general relativity, teaching us much about it; but there are significant differences, as we shall see.

The previous section showed that relative simultaneity behaves even more differently than the intuitive idea of the present than we might have expected. As a reference body accelerates, events move back and forth between (relative) past, present, and future. (And this fact doesn't depend on accelerating in some unusually extreme way: in special relativity, if you start *walking*, there will be some distant events that change from past to future.) So, even beyond its relativity, the relative present does not behave at all like an intuitive conception of the present.

In general relativity, the concept of the present is even weaker; because spacetime can be *curved*, there is not even any special way of slicing it into relative instants.

In chapter 7 we contrasted curved *space* with flat Euclidean space. Now we analogously contrast curved *spacetime* with 'flat,' Minkowski geometry. A detailed explanation of what this means is beyond the scope of this book (see Further Readings in chapter 1 to get started), but the idea is intuitive enough for our purposes.

Think about how we might describe a relative present in a curved spacetime; for simplicity, picture one time and just two space dimensions. With one spatial dimension we have drawn *lines* of simultaneity by reflecting a body's worldline in photon worldlines. In two spatial dimensions we generalize that rule to give a *surface* of events that are simultaneous relative to the reference body. The rule is straightforward, but the details aren't important for us. (For those with some mathematical background: in each direction, draw a straight line such that a photon worldline bisects the angle between it and the 'component' of the worldline in that direction.) In flat, Minkowski spacetime the surfaces behave like the lines we have been drawing; indeed, they are the planes you get by projecting these lines at right angles to the page.

But in a curved spacetime, this rule may give a surface that is not purely spatial. There can be paths *entirely in the surface* along which a person could travel. But the end of a journey has to come later than the start, so the surface must be extended in time as well as space. (We'll return to this example later.)

In other words, the familiar definition of a relative present fails in general relativity, and as far as I know no one has suggested an alternative that fares any better. It's not that the definitions don't work in any spacetime, it's that they don't work in all the possible spacetimes of general relativity. Metaphorically, think up any definition of a relative present, and you will find books in the 'library' of general relativity in which it breaks down. So the present is not a fundamental notion of the theory, rather a structure that arises if the geometry is just right (if it's flat, for instance).

That doesn't mean that everything we have said is irrelevant to general relativity. Curved geometries look flat in small enough regions around any point, as the Earth in our vicinity is treated as flat. So in curved spacetime, in any sufficiently small region, spacetime is approximately flat, so the relative present makes sense, and so does everything we saw following from it—that clock and bodies in relative motion run slow and shrink, for instance. That's why high-energy physics labs, extending over miles, use special, not general, relativity.

15.3 TIME VERSUS SPACE YET AGAIN

In our earlier discussion of time, the present played a central role. First, when we constructed the block, we stacked up purely spatial slices, all the 'presents' that make up time. Then we considered the roles that the present plays in our conceptions of time and change. But now we have seen that relativity plays havoc with our intuitive notion: the present is at best relative, that conception allows events to move back and forth from past to future, and ultimately it's not even a basic notion at all. This last point suggests that we should give up thinking of our relative 'present' as really capturing anything fundamental about time at all. (Though it still makes approximate sense when spacetime is flat enough.)

But if we give up the present, then we give up our intuitive way of distinguishing time and space: time as a series of spatial presents. So how do they amount to distinct components at all? Is the distinction obliterated? No, but the spatial and temporal aspects of spacetime are distinguished in a very new way, as we shall now see. We'll start in flat spacetime, then generalize.

Nonrelativistically, picking a single event—the Gettysburg Address, say—picks out a moment in time—all the other events that occurred simultaneously—which divides spacetime into the past and present of the event. Of course in special relativity the phrase '*the* simultaneous events' has no absolute meaning, for relative to different reference bodies, different events will be simultaneous. However, there is an absolute answer to the question of whether there is some possible reference body relative to which a pair of events are simultaneous: 'yes' as long as they lie on a line of simultaneity of some possible worldline, otherwise 'no'.

The answer depends, then, on which worldlines are possible: *those corresponding to motions slower than the speed of light*. We don't have room for a complete explanation, but here is one important intuitive reason why 3×10^8 m/s acts as a kind of cosmic speed limit.

We can see from our Minkowski diagrams that if something is traveling slower than the speed of light relative to one reference body, then it is relative to all. Lines at $45°$ represent the worldlines of photons, with a speed of 3×10^8 m/s relative to every reference body by the constancy of the speed of light. So anything with a worldline closer to the vertical than $45°$ is traveling at less than 3×10^8 m/s relative to every reference body—less distance per second.

Hence if I am moving at 99.9 percent of the speed of light relative to you, say, anything that is moving slower than light relative to me—at 99.9 percent of the speed of light, say—moves at less than 3×10^8 m/s relative to you, *not* at nearly twice the speed of light. That is, another effect of the constancy of the speed of light is that the velocity of A relative to C is not the velocity of A relative to B plus the velocity of B relative to C, and is always less than 3×10^8 (or equal to it, if A is a photon).

Thus our question of whether two events are simultaneous relative to any possible reference body is that of whether there is any possible motion slower than the speed of light for which the line of simultaneity includes both events.

So picture a point p in a Minkowski diagram, all the worldlines through it that are closer to vertical than $45°$, and all the corresponding lines of simultaneity–according to our rule, *every line between $45°$ above and below horizontal*. From what we have said, all and only points on those lines are simultaneous to p, relative to some possible body. See figure 15.2.

Our discussion has shown that relativity divides spacetime up into three regions relative to any point—the 'here-now'. First there is a region

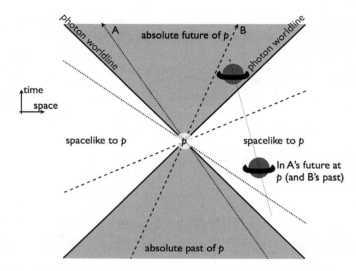

Figure 15.2 'Causal' Structure: Pick any point 'spacelike' (unshaded) to
p—'here-now'—and it is on the line of simultaneity of some worldline
slower than light: relative to someone it is simultaneous with *p*. But pick
any event in the 'absolute past' (or future), and there is no such
worldline.

of points that are simultaneous with here-now for some relative present.
Since, relative to some reference body, these points are displaced in space
only, they are called 'spacelike separated' from here-now. Second, there
are the points above the spacelike region. Since all of these points are
in the future of all possible lines of simultaneity through here-now, this
region is called the 'absolute future'. Finally, there is the similar region
below—the 'absolute past'.

There's another way to think of this division. Since nothing can travel
faster than light, the absolute past contains all occurrences in the block
that could have any effect here-now. For instance since there is a cup of
tea here-now, it must have been poured in the absolute past of here-now;
otherwise it would have to have traveled faster than light to get here! Or
again, the light reaching me now from the sun must have originated from
just outside the region spacelike to here-now; after all, it traveled here at
the speed of light.

Similarly, the only events that one can affect are those in the absolute
future of here-now. Light takes over an hour to travel between the Earth
and Saturn, so nothing can reach there in less time. At present the Cassini
spacecraft is in Saturn's orbit, so suppose *p* in figure 15.2 represents
mission control's location (NASA's Jet Propulsion Lab) and time (now);
imagine that they want to instruct Cassini to take a picture. Some of
Saturn's worldline is also shown, including a point spacelike to *p*. Even
though it is in the future of *p* relative to some worldlines (e.g., A), it

is impossible to transmit an instruction at p that will reach Cassini at that point; the message would have to travel faster than light. (And note that this fact depends only on the relations between the points; it is not motion-relative, and the impossibility applies equally to A and B.) A command sent from JPL now (i.e., from p) can reach Cassini only at some point in the absolute future.

In other words, all and only events that could possibly affect me are in my absolute past; and all and only events on which I can have any effect are in my absolute future. In this regard, the *absolute* past and future are like the prerelativistic understanding of the past and future. (Conversely, the *relative* past and future are unlike the prerelativistic concepts: for example, it is impossible to affect any event in the relative future that is also spacelike to here-now). Because of this understanding, the division illustrated in figure 15.2 is often called the 'causal' structure of spacetime: it describes which points can affect which. (It is also called the 'lightcone' structure: imagine a second spatial dimension out of the page and photon worldlines at $45°$ in all directions; you get *cones* above and below p, representing circles of light converging on p and expanding out from p.) It captures the deepest sense in which space and time are distinguished in relativity.

The causal structure, unlike the relative present, is still a deep feature of general relativity, but we have to be careful thinking about its significance. To see why, let's use it to illustrate a little of what it means for *spacetime*, not just space, to be curved. 'Lightcones', like those illustrated in figure 15.2, depend on both the spatial and temporal aspects of spacetime and so are sensitive to the way they curve together. Broadly, the lightcones 'tip' relative to each other, as the spatial and temporal aspects of spacetime are 'tilted' around by its curvature.

For example, consider the spacetime geometry illustrated in figure 15.3, 'discovered' (in a mathematical sense, not by astronomy!) by the logician Kurt Gödel. The horizontal plane represents the surface discussed in section 15.2 as a possible generalization of the relative present to general relativity. As we follow a radial line out, the lightcones tip ever further *until some lines in the surface are inside the lightcone*! Hence some points are in the absolute futures of others; as was said before, the surface doesn't make sense as even a relative present.

But there's more! The tipping happens in the same way in every direction: imagine the radial line rotated about the vertical. That means that there is a circle in the plane that runs to the future of every point through which it passes; it is a worldline that is everywhere slower than light. So it is a worldline a body could possibly follow, which, because it is a circle, returns to its starting point: it represents a journey back to to the very same point of space *and time* where (*and when*) you started! And of course, on the way, through *earlier* points of time. Gödel's spacetime shows one way that time travel is possible according to general relativity.

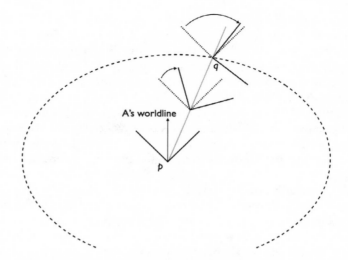

Figure 15.3 Gödel Spacetime: the lightcones along *pq* are increasingly tipped away from that at *p* by the curvature of spacetime (conversely, along *qp* from that at *q*). The lightcone at *q* is so tipped that curves in the horizontal plane pass into the absolute future of *q*.

It also shows that care is needed in understanding the causal structure. In this example, some points are in the causal past *and* future, so the relativistic causal structure does not distinguish past and future in the intuitive way. Absent any other structure, general relativity simply does not make such intuitive distinctions; they are possible only if the geometry is just so. It's our best theory of space and time, so we've discovered that, according to our current knowledge, the intuitive distinctions are not fundamental physical ones.

Note, however, that the lightcone does divide spacetime in such a way that no points are both past and future, in a region around each point. In figure 15.3, consider a region around *q*, too small to contain the loop back. So we can say that sufficiently *locally* the causal structure does divide spacetime up into past and present.

We will leave things there—a place to start further inquiries—except to make a couple of remarks. First, there are other pieces of physics that may be relevant here: for instance, there are quantum mechanical theories that introduce an absolute standard of rest, and there are possible universes in general relativity in which there is a natural, 'absolute' present.

Second, we should ask briefly where this chapter leaves the nowist (if he still wasn't convinced by the arguments of the previous chapter). Well, all of the problems discussed before are still in effect; the question is how he responds to the new picture we have developed. He could deny that the generally relativistic account of time settles the issue and, say, postulate an absolute present. Then he owes either strong scientific reasons for so doing (as quantum mechanics might provide) or strong reasons for giving up on physics as the right way to investigate time. Alternatively, the nowist could appropriate some structure of general relativity to replace the present. For instance, perhaps what moves is not the present but the spacelike region. Of course, that region is defined only relative to a point—the here-now—so its supposed motion would have to be relative to a series of points: a worldline. And if nowism responds to McTaggart by restricting existence to the 'present', then we find spacetime made up of a pattern of overlapping realities, each relative to a point.

15.4 EINSTEIN'S REVOLUTION IN PHILOSOPHY

We've seen repeatedly in this book how philosophy can learn from physics, how philosophical issues can be informed or even settled by the best physical theories at hand. Of course, the point of the book was to show that physics and philosophy, though very different in some ways, can be in fruitful dialogue. Relativity makes the point very dramatically.

In our earlier discussions of the philosophy of space and time we assumed a very intuitive picture in which we could divide the block universe up neatly into space and time. That is, we could sensibly speak of 'an instant' and pick out a slice of spacetime unambiguously. What could be more obvious than that? Didn't it seem that such an idea followed from the very idea of space and time? Didn't our picture seem like something to be assumed by any science, as a framework in which more detailed laws of motion and forces might be described?

It might have seemed that way, but we have seen that physics, based on a rock solid experimental foundation, shows that that picture is just wrong, because the present is relative (and ultimately not fundamental). And so are many other things that go with it: relatively moving bodies contract and relatively moving clocks slow down, speeds don't add up, and trips along different paths will set clocks out of sync. Any philosophical discussion that involves space and time has to be reconsidered to see whether it uses the concepts correctly in the light of our new knowledge.

If our knowledge of something as basic as space and time can be revolutionized, we have to ask what else might be changed in the future. Things that seem like harmless assumptions now may also be shown to be wrong by future science, and philosophy will have to take such changes into account.

I was careful in our earlier discussion not to make claims that would be rendered false by relativity, and to stick to topics that would remain relevant. In this chapter we've discussed how those debates are affected. But let me mention briefly a philosophical debate that, perhaps surprisingly, must take relativity into account.

Where is your mind? Well, if it's your brain, then it's in your head. But on the other hand, perhaps the mind is immaterial, or perhaps not, but we cannot simply identify it with your brain for some other reason. Then where is it? Faced with this puzzle, some philosophers have suggested that it is not in space at all; it's nowhere. Descartes held that the mind was non-spatial, a view something like this, and in the twentieth century Ludwig Wittgenstein also argued for it. On that view while mental events—our thoughts, for instance—occur at moments of time, they don't occur at points of space.

But as my professor Robert Weingard pointed out, it's hard to make sense of such a view given special relativity. For it entails that mental events can be separated in time while not being separated in space; they occur at different times, but don't occur at any places. But, as we have seen, events separated in time relative to one reference body are separated in time *and space* for others in relative motion.

Perhaps relativity isn't strictly incompatible with the view in question, but at the very least some kind of new story will have to be told. The point is that physics can affect philosophical questions in areas seemingly quite remote.

Further Readings

For philosophical reflections on special relativity I recommend *Time and Space* by Barry Dainton (McGill-Queen's Press, 2001)—especially for further discussion of nowism—and Brain Greene's *Fabric of the Universe* (Knopf, 2004)—especially for the problem quantum mechanics causes for the relativity of simultaneity.

Weingard's paper "Relativity and the Spatiality of Mental Events" appeared in *Philosophical Studies* in 1977 (279–284).

16

Hands and Mirrors

Philosophers of physics are of course interested in some of the most fundamental concepts of physics; we've emphasized space, time, and motion. This chapter will discuss another aspect of space: the property of being 'handed'. That is, some things—such as screws, scissors, and hands—come in two varieties, which we call 'left' and 'right'. It's pretty clear that these are spatial properties, but what are they? We'll see how this question has implications for our discussion of the nature of space in chapter 9. I will argue for a relational account of handedness.

Some of our discussions have relied on some quite specific physical theories: especially Newtonian mechanics and relativity. Others have relied on even more basic ideas, such as topology and the calculus to help us understand basic concepts such as shape and motion. Our discussion here is of the latter kind; I want to know in the first place what it means to be left or right, and that is largely a geometric question.

However, the issue is, like those of motion and shape, relevant to more specific physical theories. First, according to our current understanding of particle physics, the weak force (discussed in chapter 10) acts differently on left- and right-handed systems. I won't consider here what that means (see Gardner in the Further Readings), but our understanding left *versus* right is a first step in that direction. Second, as we shall discuss at the end, handedness depends not only on handed things but on the topology of space. Thus the experimental work into the shape of space that we discussed in chapter 4 has a bearing on our current investigation.

16.1 IS HANDEDNESS INTRINSIC OR EXTRINSIC?

Let's start with an observation of Kant's:

> It is apparent from the ordinary example of two hands that the shape of one body may be perfectly similar to the shape of the other, and the magnitudes of their extensions may be exactly equal, and yet there may remain an inner difference between the two, this difference consisting in the fact, namely, that the surface which encloses the one cannot possibly enclose the other.

Think about your hands (assuming that you have the standard issue). Ignoring scars, warts, and other minor differences, they are very much alike. In particular they have the same volume, the same surface area, the same width palms, and the same lengths of fingers and thumbs. Specifically, they are geometrically 'similar' and of the same size. But they are also crucially different; they will not fit in the same space. So, a left hand will fit in a left-handed glove but a right hand will not (without stretching the glove into a new shape). We say that they are *incongruent*; conversely, two bodies are *congruent* if they will fit into the same space.

We recognize the similarity but incongruence of hands by saying that they are *handed* or *chiral* (pronounced with a hard *k*), and learning that they come in two types of 'handedness' or 'chirality', called 'left' and 'right'. Contrast this situation with, say, basketballs, or bricks, or pyramids: any two spheres or two cuboids or two tetrahedra that are the same size will fit into the same spaces, and so are congruent, and so do not come in two chiralities.

What's the difference between hands and the other solids? Spheres and cubes and tetrahedra have 'planes of mirror symmetry'. If they are reflected in these planes, the result is an identical solid in the same place: if a sphere is reflected in a plane through its center, then the result is an identical sphere. It follows (assuming Euclidean space, an important assumption to which we shall return) that *any* mirror image of a sphere (or cube or tetrahedron) is congruent to it.

Hands, however, do not have any planes of symmetry. Hold your right hand up in front of you, palm away: your fingers make up the top half, which are not the mirror image of the base; the left side contains the thumb, which is not the mirror image of the pinky; and the palm is not the reflection of the back. Reflections of a hand, then, need not be and indeed are not congruent to it.

So there's no great mystery about what it is to be handed: it's to lack a plane of mirror symmetry. Thus we understand why a variety of other things are handed: gloves, the spirals on shells, screws, helical molecules, a page of writing, the face of an analog watch, and so on; see figure 16.1. (Actually, that's not quite the whole story: while objects with planes of symmetry always have congruent mirror images, objects without planes of symmetry may or may not have incongruent reflections. We'll skip the rest of the story for simplicity: see chapter 3 of Gardner.)

But there's another question, which Kant raises in the essay quoted above. What is it for an object to be of one handedness rather than the other: for example, what is it for my left hand to be *left* rather than *right* (or for a screw, spiral, or molecule to be right-handed and not left-handed)? According to Kant, the property is an *intrinsic* one, an 'inner difference', not an *extrinsic* one.

Intuitively, the difference is that something's extrinsic properties all depend on relations to other things, and on their properties. So, for

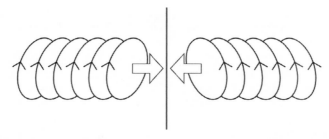

Figure 16.1 Two mirror image spirals; clearly they are incongruent.
Lots of things have basically the same basic shape and so are handed:
screws, certain organic molecules (the double helix), and snail shells, for
example. Suppose the spirals represent screws, then the big arrows show
that they move in opposite directions when turned in the same sense.

instance, the properties of presently touching a keyboard or of being
bigger than my sons' right hands are extrinsic properties of my hand,
for they depend on the distance between my hand and the keyboard and
on the size of my sons' hands. On the other hand (!), the mass or shape
of my hand is intrinsic, apparently independent of other things. (Or is it?
Would my hand have the same mass if everything else suddenly halved
in mass, or would we say it had doubled its mass and everything else
stayed the same? And it couldn't have the shape it does if the geometry of
space were different. The intrinsic–extrinsic distinction is philosophically
contentious, but we will short-cut around this controversy by considering
some specific proposals for handedness for which the intuitive distinction
works well enough.)

Against Kant, I'm going to defend the claim that being of a particular
handedness—being left rather than right, say—is extrinsic. If you think
back to chapter 9 and my argument for a relational view of space, you'll
see why: left and right hands are exactly alike in the spatial relations that
are intrinsic to them. Left and right hands (or left and right anything) have
exactly the same relational descriptions: assuming the hands are held the
same way, the distance from tip to bottom of the index finger is the same,
the distance from thumb knuckle to pinky knuckle is the same, and so
on for any two points on each hand. If you were given a list of all the
relations between the parts of a hand and were asked to build an object
that realized those relations, you could build either a left or right hand,
and both would follow the instructions exactly.

Since the chiralities of the hands differ but the relations of their parts
do not, their handednesses cannot be facts about those relations. But if,
like me, one accepts the relational account of space, then the relations of
the parts are the *only* spatial properties of the hands that are intrinsic to
them, and so having a handedness cannot be an intrinsic spatial property.
Since having a handedness is a spatial property of some kind, it must

be extrinsic: it must be a property of the spatial relations that a handed object bears to other things. Just what property we shall discuss in the next section.

16.2 THE 'FITTING' ACCOUNT

Imagine a world in which the only things in existence are similar and equally sized hands, some of which are incongruent. These hands are naturally divided into two types: all the hands congruent to each other are of one type, and all the others—which are congruent to each other but incongruent to any hands of the first type—are of the other. One of these types we can call 'left-handed' and one 'right-handed,' but which is which?

The answer is that what we call each collection of hands is a matter of linguistic accident or convention. Just as we could have used the word 'foot' to name our heads and the word 'head' for our feet, we could have called right hands 'left' and vice versa. As conventions, these choices don't show anything about the world. We wouldn't have feet for heads if we called our heads 'feet', and similarly, that we call one collection of hands 'left' doesn't show anything interesting about handedness. Certainly, if I started calling my left hand 'right', it wouldn't become a right hand.

So when we ask of some hand in our imaginary world what it is for it to be left handed, say, we are just asking why it belongs to the collection of hands that we have happened to call 'left'. The answer is simple: it belongs to that type because it is congruent to all the hands in the corresponding collection. Thus, being of one chirality rather than the other is a matter of being congruent to one collection of hands rather than another.

Since congruence to other hands is an extrinsic property—whether a hand is congruent to another depends on the shape of the second hand—it follows that in the imagined world, *being of a particular chirality is an extrinsic property.*

Now suppose that there are some other things that lack a plane of symmetry: (1) some smaller hands, (2) some fists the same size as the original hands, (3) some gloves of the same size, and (4) some screws. These things are all handed, and so each of the new types of things can be divided into two kinds by (in)congruence: two kinds of fists, for example, so that fists of the same kind are congruent and fists of different kinds are incongruent.

What, now, is it for an object to be of some chirality: left rather than right, say? For instance, a fist's being congruent to the members of one collection of fists rather than another can't now be enough for it to be left. The fist must also be of the same handedness as one of the collections of original hands, as one of the collections of small hands, as one of the collections of gloves, and as one of the collections of screws, but fists are incongruent to all those things. In other words, we also need to say which collections of small hands, fists, gloves and screws have the same handedness as the collection of hands that we have dubbed 'left'.

Here's how to do so, again relying on extrinsic properties:

(i) First, any two hands have the same handedness if they can be made congruent by a change of size alone: thus a small hand is a left hand if—were it to expand uniformly, then it would become congruent to the original left hands.

(ii) Next, any fist and hand have the same handedness if they can be made congruent by changes of size and motions within the 'normal human range of motion for hands' (such as making or unmaking a fist).

(iii) A hand and a glove have the same handedness if the outer surface of the hand and the inner surface of the glove can be made congruent by changes of size, normal human motions, and glove movements that don't leave the glove (badly) stretched.

(iv) Finally, a hand and a screw have the same handedness if when the fingers of the hand are curled into the palm and the screw turned in the sense that the fingers point, the threads make the screw move in the direction of the thumb.

This (extrinsic) account of handedness, then, has three parts: first we note that each sort of handed object is automatically divided into two kinds, or handednesses, by the extrinsic property of congruence. Second, we have to say for two different sorts of object which handednesses are the same: which collection of congruent hands has the same handedness as which collection of congruent gloves, say. In each case, we have done this by reference to extrinsic properties, either by congruence after some change or, in the case of screws, by the property of moving in a certain way relative to a hand when turned in a certain way relative to it (it's because the motion is judged *relative to the hand* that this is an extrinsic property of the screw).

Finally, once all the dividing into kinds and matching of kinds is done, once the two handednesses are distinguished for objects of any kind, then we can choose which handedness is to be called 'left' and which 'right'. But it is clear from our account that the properties these words name merely amount to being in one collection of objects rather than another. Thus being of one handedness rather than another depends on the extrinsic properties that actually distinguish the two collections. (Note too that the properties invoked involve only spatial relations, so the extrinsic account makes sense of handedness for the relationist, putting to rest the worry that I raised at the end of section 16.1.)

In each case, the judgment about same handedness roughly amounts to a rule about how things 'fit' together in an agreed sense: two similar hands have the same handedness if they fit together in the sense of congruence;

a hand and glove have the same handedness if the hand fits the glove; a fist and hand have the same handedness if the fist fits the same space as the hand when unclenched; and even a screw and hand have the same handedness if the threads and curled fingers 'fit each other' in a loose sense. Hence I'll call what we have developed the *fitting account* of handedness.

Of course, the story is somewhat contrived in order to make transparent how one can understand handedness, but if you think about it, the fitting account is basically the correct account of how we do learn and make judgments about left and right. We in fact do make judgments about whether the surfaces of a hand and a glove can be brought into coincidence by sliding the hand into the glove, actually or mentally. And when we learn to distinguish the left-handed variety of some object, we have to learn what relations it has to existing objects, and in fact the relations in question often are relations to our hands. For instance, you may well have actually learned the difference between left and right screws by the rule that I gave earlier.

The examples also indicate a couple of interesting features of handedness. First, while the definitions of same handedness for the various hands and for the various gloves seem completely natural, when it comes to screws things seem a bit more arbitrary: why not say instead that the screw is right-handed if it moves forward when turned in the tip-to-base sense of the fingers on my right hand? Or suppose we discovered that the world contained a great many irregular but geometrically similar rocks (imagine a lot of identical spiky stones) with no plane of symmetry. We might come to distinguish some as having the same handedness as our right hands, but their irregularity makes the choice of how to do so entirely arbitrary—we could equally well have identified those rocks as the left-handed ones.

What this arbitrariness shows is that there is no notion of 'same handedness' intelligible independently of reference to specific bodies; the various definitions of same handedness that we gave are not instances of some prior general definition of same handedness, applicable to all handed objects, but always extensions of the concept of handedness. Even when the way to extend the concept is obvious—of course hands of different sizes have the same handedness if they would be congruent after a change of size—there is nothing in the prior rules of same handedness that forces the obvious extension. What it means for a new kind of handed object to fit existing ones is at most guided and is always ultimately a free choice (like the initial choice of what word will denote which group of congruent objects).

The second feature of handedness illuminated here is that the definitions of left and right may be vague. What exactly is and is not within the 'normal human range of motion'? When is a glove 'badly' stretched? I don't see any problem here. It's just a fact, regarding many real-world objects which we divide into left and right, that there are borderline cases.

For example, after a while some cheap gloves might become so stretched that it's hard to distinguish left from right, and you might even make different assessments on different occasions. All vagueness means is that in everyday judgments about handedness, we don't always use geometrically precise notions.

16.3 KANT'S ARGUMENT AGAINST THE FITTING ACCOUNT

It would thus be hard to dispute that our practical understanding of handedness is extrinsic, but one could argue for some additional 'deeper truth' about handedness according to which left and right were intrinsic properties. Indeed, Kant himself held such a view, as the quotation in section 16.1 suggests. We'll how look at his argument, and I will explain why it fails, why the fitting account describes not only our everyday understanding of handedness but everything about what it is to be left- (or right-) handed—there is no 'deeper truth'.

Kant relied on a thought experiment. Imagine a universe in which the only thing is a hand. He claimed it surely must be either a left hand or a right hand. But since there are no other things, either way the hand must have all the same extrinsic properties; it is the largest (and smallest) object in the universe, for instance. The hand simply cannot have different relations to other objects if it's right than if it's left, because there are no other objects. And so, he concluded, the difference between being left and being right for the lone hand is not extrinsic; it is intrinsic.

But this objection is not fatal to the fitting account. The logical conclusion of the fitting theory is that in such a universe, since there is nothing for the hand to fit to, it is *neither left nor right*; the question of its rightness and leftness does not come up. (Or more precisely, all that can be said is that the hand 'fits itself', and so it has the same handedness as itself, which we could call 'left' if we wanted. It's not so much that handedness fails to apply, but that it is utterly trivial.)

Further, if one accepts the fitting theory, then one cannot ask whether the lone hand has the same handedness as *my* left hand. Being in a different, imaginary universe, the lone hand can't be moved into the same region of space as either my left or right hand, and so the question of its (in)congruence to them makes no sense.

And so, according to the fitting theory, the single hand, alone in the universe is neither left nor right, either with respect to anything in its world or with respect to anything in our world. Instead of conceding that the handedness of the lone hand would be intrinsic, the fitting theorist says that it simply has no handedness.

Kant foresaw this response and tried to prove that the lone hand *must* have a handedness. Suppose (taking a few liberties with his argument) that an idealized human body, whose left and right sides are exact mirror images, were now created alongside the lone hand. The single hand must have the same handedness as exactly one of the body's hands, but which? Kant believed that the only way to answer that question would be if the lone hand were left or right *before* the body arrived: if left, say, then it would have the same handedness as the body's left hand.

Must the fitting theorist concede that the lone hand is handed (and so concede a 'deeper' intrinsic truth about handedness)? Only if he must accept that there are *two* possible outcomes to the introduction of the body: being the same handedness as one hand or being the same handedness as the other. Otherwise there is no need for the lone hand already to have one handedness or the other before the body appears.

But the fitting theorist does not need to accept that there are two possible results of adding the body. In whatever way the body is introduced, the lone hand has the same handedness as one of the body's hands and the opposite handedness to the other. In other words, it fits one of the body's hands and not the other, and so the division of hands into two kinds of handedness leaves us with two hands of one type and one of the other. We can call the type with two hands either 'left' or 'right', but that's just a linguistic convention, not corresponding to two different results of introducing the body (just as there aren't different outcomes depending on whether we decided to name the body Wilber or Yves). Thus the fitting theorist will say that there is only one possible result of introducing the body, not two, and so no need to suppose that the lone hand was already handed. As you should expect from an extrinsic account, the hand takes on a handedness only when there are some other hands around.

The fitting theorist's retelling of the story reveals a fallacy in Kant's argument. He concludes that there are two possible results of the creation of the body, because he supposes that the lone hand can end up as the same handedness as either the left or right hand of the body. On this supposition, one reasonably concludes that which outcome arose depended on which handedness the lone hand had before the body arrived.

But Kant's supposition itself presupposes that the body's hands have their handednesses as intrinsic properties; because the body has a plane of mirror symmetry, they have all their extrinsic properties in common. Obviously, the fitting theorist will not accept this presupposition. Since Kant is trying to argue against the extrinsic account, his reasoning is circular if he assumes that handedness is intrinsic at any point. It amounts to saying that handedness is intrinsic because it's intrinsic, which is no reason to believe the theory at all.

(Philosophers say that when one assumes what one wants to argue for, one 'begs the question'. Unfortunately—probably because of half-remembered philosophy classes—this phrase is often used to mean that some claim 'raises a question', as the truth of global warming raises the question of what we should do about it. It puts philosophers' teeth on edge to hear the misuse of 'question begging', though it's probably our own fault for not making our students pay better attention.)

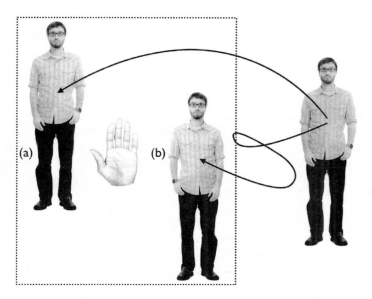

Figure 16.2 If a nonsymmetric body is introduced into a lone hand world, there are two possibilities: here so that the lone hand is the same handedness as (a) the watch-wearing hand or (b) the other hand. But according to the fitting theorist there are two possibilities simply because it is not specified how to insert the body relative to the hand: 'flipped over' as in (b) or not as in (a).

Kant's original argument may fail, but there are some ways that one might try to modify it. Suppose that the body is in some way nonsymmetric; there is a wart or a watch on one hand, or one hand is just heavier than the other, say. Or suppose that the body is an exact copy of someone in our world, whose hands are left and right according to the fitting account, and that we can identify one of the hands of the created body as the copy of the left hand of its original (and similarly for the right hand)—that the created body's hands 'inherit' the properties of left and right from the original.

Then, as you see from figure 16.2, there will be two ways to introduce the body: so that the lone hand has the same handedness as, say, the warty or the watch-wearing hand and not as the the unwarty or watchless hand; or the same handedness as the copy of the left hand and not as the copy of the right hand. Then we have to find some explanation of why two different things can happen when the body is created. The body itself is just the same either way, so are there two ways for the lone hand to be ahead of time, left or right, one of which leads to one outcome and the other to the other?

The fitting theorist cannot accept any such initial difference, so how does he understand the two possibilities? The reasonable reply is that there are two because the thought experiment is not fully described. In particular, given that the hands of the body are distinguished, there is simply a choice about how to insert the body in the world relative to the hand, related as mirror images.

Analogously, as far as the story specifies, there are a lot of possible distances from the hand at which the body could be created, but clearly this ambiguity arises because the story leaves some things vague. Certainly, we don't feel any need to introduce a new intrinsic property of 'far apartness' for the original hand, one that fixes where the body will be created. The fitting theorist will say that the case is just the same for handedness: there's more than one possible outcome because the story leaves details out, not because of an intrinsic property of leftness, say, of the hand.

Summing up, there is nothing in Kant's argument to worry the fitting theorist: the extrinsic account of handedness makes perfect sense of all the thought experiments. On the other side we have the intelligibility of the fitting account, and the good job it does of making sense of our actual judgments of handedness. In my book (and so in this book!), the fitting theory is the clear winner.

16.4 LOOKING LEFT AND RIGHT

There might seem to be something missing from the fitting account that could yet undermine it. When I see my left hand, I see that it has the same handedness as my left glove and foot, and the opposite handedness to my right hand and normal screws, in accordance with the fitting theory. But in addition I see that it is my *left* hand, which seems to be something additional. And so the question is what extra is needed in order for the mind to perform this judgment: what does it take for me to tell not only that this hand fits some things and not others, but that this hand is in fact left and not right? If the answer is not compatible with the fitting account, then that account would not be complete after all: maybe I see that the hand possesses an intrinsic property of leftness.

The simplest response is to point out that my hands are not exact mirror images at all and so in fact are distinguishable by their shapes. For instance, I can tell which hand is my left because it's the one with the ring. I don't find this answer ultimately satisfying: I can still tell my left hand from my right when I take off my ring, and indeed I believe I still could if my hands were perfect mirror images. But how, according to the fitting theory, is such an ability possible?

The answer is that it takes a brain state that is not mirror symmetric. Let me explain why. Consider the following thought experiment involving someone whose body is perfectly mirror symmetric, inside and out. In this experiment we shine a light on one of his hands and ask him to think to himself whether it is the left or right hand. Because of the mirror symmetry, at the start of the experiment the brain state of the subject is symmetric about the plane dividing the body: the brain itself is symmetric about the plane, and any properties of the brain (for instance, that certain neurons are firing) are also distributed symmetrically (so that the mirror-image neurons fire simultaneously).

First we shine the light on the left hand (we certainly know left from right, and we imagine ourselves in the laboratory shining the light, so *we* have no problem telling which is the subject's left). The subject's brain state will evolve in a certain way according to the laws of physics—it is a physical object—depending on its initial state and on which hand is illuminated, and so will his mental state, until, let's suppose, he correctly thinks, 'aha, it's left'.

Now we repeat the experiment, starting the subject in the same symmetric initial brain state; our question is whether *that* state could correspond to a state of knowledge of which hand is left and which right. Now because the brain is symmetric, the conditions at the start of the second experiment are a perfect mirror image of the conditions at the start of the first. It follows that this time the evolution of the brain will be a perfect mirror image of the evolution of the first, so that the brain will end up as a mirror image of itself at the end of the first experiment. (There is a substantive assumption here that the laws of physics governing the brain in this case are mirror symmetric and don't involve chance; this assumption seems plausible given contemporary physics, and given that we can reliably make the judgment.)

Suppose that the state of the brain is determined by its intrinsic properties. Since reflections do not change the intrinsic properties (according to the fitting account), the subject's brain thus ends up in the same state in both experiments. And suppose further, as we have discussed before, that the state of a person's mind is determined by the state of his brain, so that the person ends up in the same mental state in both experiments: the state of thinking, 'aha, it's left'. But in the second experiment this is the wrong answer! And so we see that the subject, because his brain was symmetric, does not really know the handedness of his hands at all. Indeed, given his responses, his initial mental state seems to correspond to the knowledge that 'left' means illuminated (or picked out in some way) when applied to his hands.

However, if our brain states are asymmetrical, then we can do the job. To see how asymmetry defeats these kinds of considerations, we don't need a realistic account of how the brain is configured. Any asymmetry will do. So imagine a new subject with a different initial brain state, one in which the right side of her brain contains a discernible R-shaped part and the left a discernible L-shaped part. Now the initial state when her left hand is illuminated is not the mirror image of the initial state of when her right hand is illuminated; the L and the R are at different distances from the illuminated hand in the two cases. And so the brain states at the end of the two experiments will not be mirror images, and so they will not (necessarily) correspond to the same mental states.

Thus there seems no reason to think that the fitting theorist cannot account for all the facts concerning our ability to know which hand is left and which is right: the brain state described will, in principle, do the trick. The fitting account is all we need to understand handedness.

One last consequence of this analysis. Consider someone who is the perfect mirror image—of all parts, internal and external—of our subject with the asymmetric brain. Her L-shaped region will be on the right and her R-shaped region on the left of her brain (and of course are reflected L- and R-shaped). If we now illuminate this mirror image's right hand, her initial condition will be the exact mirror image of the initial condition of the original asymmetric subject when her *left* hand was illuminated. And so by the reasoning used earlier, our new subject will end up thinking incorrectly 'aha, it's left'. Similarly if we illuminate her left hand, she will end up thinking 'aha, it's right'. And so, if Earth is ever invaded by our duplicates from a mirror image universe, we will be able to tell our friends from our foes by testing whether they can correctly identify their left hands.

16.5 MIRRORS

There is one final puzzle concerning handedness that the fitting account should be able to solve: why do mirrors reflect left–right but not up–down? When you look in a mirror, there appears to be a second body, on the other side the mirror, which is the mirror image of yours (of course it is only the 'appearance' of another body; no one is really there). The left side of this mirror image faces your right side, but his or her head is level with your head, not your feet.

What is the source of this asymmetry? Not the way mirrors work: the effect is the same if you turn the mirror through 90 degrees! One might suggest that it is because left and right sides possess intrinsic properties of leftness and rightness, which are switched (in appearance) by mirrors, but that heads and feet have no analogous properties of 'up-ness' and 'down-ness' to be switched. We'd better have an explanation in terms of the fitting account to avoid this suggestion, since we claim that left and right are extrinsic.

A first response is to wonder whether the question isn't simply confused. My feet and head are just not mirror images, so there's no reason for me to expect to see my feet reflected opposite my head and my head reflected opposite my feet, while my hands are mirror images of each other so of course the reflection of my right hand is a left hand and vice versa. To think that the reflection of my head could be a pair of feet would just be to misunderstand what a reflection is.

But this response is not enough. It tells me that I should (of course) expect the reflection of my left hand to be a right hand and that of my head to be a head, but it doesn't help us understand why the mirror image of my *left* hand appears on the *right* side of my mirror image, while the mirror image of my *head* appears on the *top* of my reflection. This puzzle depends on how we judge which sides of my mirror image—the body that appears to be on the other side of the mirror—are right and left and which ends are top and bottom.

Think of yourself standing in front of a full-length mirror and imagine moving around the back of it to place yourself where your image appears; pretend that it is a real second body standing there facing you. Likely you imagine moving so that your right hand is now where the mirror image of your left hand is—hence left becomes right—and your head where that of your image is located—hence up and down are not exchanged. You are of course judging left–right and up–down by mentally 'fitting' the mirror image to your own body.

What is the source of the asymmetry between left–right and up–down in this thought experiment? It is the way that you moved your body around the mirror, effectively rotating yourself about a *vertical* line in the mirror, around the sides. So now repeat the thought experiment but rotating your body about a *horizontal* line in the mirror, jumping over the top, head first. Now you are upside down with your head at the feet of your reflection (and vice versa) with your left side on the left side of your reflection (and vice versa).

Why isn't this the right way to 'fit' your body to its mirror image? It would imply that mirrors reflect up–down and not left–right in an important sense (though of course the mirror image of your left hand is a right hand and that of your feet is not a head!). It seems wrong, but why exactly?

Is it something to do with the mirror? But a mirror has the same effect however it is rotated in its own plane, so mirrors have no preferred orientation, no preferred axis, in the plane. Therefore the mirror itself doesn't determine how to compare your body to its reflection. Is it then perhaps gravity and our modes of movement that make us think this way, because we find it so much easier to walk around things than to leap over them head first? I don't think so.

There is, after all, a crucial geometric difference between going around *versus* going over, namely that going over does not bring the two bodies into (near) coincidence—for instance, feet and heads are not the same shape—while going around does—my feet are the same shape as those of my mirror image! So what we should say, in accord with the fitting account, is that the rule we have adopted (quite naturally of course) to judge which sides of another body are up, down, left, and right (and back and front) is to 'fit' that body to ours by imagining ourselves (approximately) coincident with that body and to transfer the names we give to our sides to its sides. And then, because our bodies have symmetry about a plane that is running back–front and up–down, but not about any of the other planes that bisect us, if you bring your bodies into coincidence with its mirror image, you will find left and right reversed, but not up and down, and not back and front.

This account explains how I can tell which of Homer Simpson's hands is his left, even though they are not in fact handed; like many cartoon characters they lack the geometric detail—knuckles, life lines, etc.—to distinguish palms from backs, and so have a plane of symmetry dividing palm from back. I can still tell which is Homer's left hand by imagining myself, roughly, in the space he occupies and seeing which of his would be coincident with my left hand: by seeing how his body fits to mine.

16.6 ORIENTABILITY

There's much, much more that can be said about handedness (as we saw, it turns out that nature does not respect reflection symmetry, so the laws of nature are 'right' handed), but we shall just briefly touch on one last item of importance. Earlier we made the assumption that space was Euclidean. This assumption was important because it turns out that objects can be handed only in spaces of certain shapes, those with the topological property of 'orientability'. In 'non-orientable' spaces, in contrast, no objects are handed!

The idea is intuitive enough. Some spaces have a 'twist' in them that make motions around the twist equivalent to reflections. The simplest example is the Möbius loop, a two-dimensional space that is like a cylinder but with its edges twisted before they are reconnected (see figure 16.3). Now, in the plane the letter 'F' has no 'plane' of symmetry, no line through it so that each side is the mirror image of the other; hence ' Ⅎ', its mirror image, is incongruent to it. However, if an F travels around the Möbius space, it comes back congruent to the Ⅎ, and so the F, and indeed any other figure, has only congruent reflections in this space.

The same thing is possible in any dimensions: it may be that even in our universe, if a left-handed glove were sent off along an appropriate path, it would come back right handed. Indeed, since the topology of our universe remains an open question, perhaps our space is nonorientable. Either way, it is a question for experiment and physics.

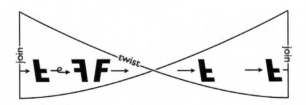

Figure 16.3 Möbius Loop

Of course the universe is a big place, and any handedness-changing path would be enormous. Hence in the context of a smaller region no objects can change from left to right just by being moved around, and so there remains a practical difference between them. So even if the universe is nonorientable, it would still make sense to classify objects as left and right on Earth, for instance.

We've now discussed handedness at considerable length and seen what we mean when we call things 'left' or 'right' and how those properties are extrinsic. We've also considered how to understand why we perceive a difference between left and right. Our story followed a pattern established in earlier chapters of considering the physical basis of our experiences; once again a philosophical question is addressed by recognizing that we are physical beings living in a physical universe.

Something similar happened when we considered mirrors, though the explanation did not rest on optics or the physics of mirrors but on the more basic geometric properties of bodies. That is, bodies composed of two mirror-image, chiral halves are coincident in just one way with their mirror images.

Further Readings

Kant's essay "Concerning the Ultimate Foundation of the Differentiation of Regions in Space" is reprinted in my *Space from Zeno to Einstein* (MIT Press, 2001). However, I quoted from a nicer translation of the same essay found in *Immanuel Kant: Theoretical Philosophy, 1755–1770*, translated by David Walford (in collaboration with Ralf Meerbote), (Cambridge University Press; 1992: 370–371).

The very best book on all things to do with handedness is Martin Gardner's *The New Ambidextrous Universe: Symmetry and Asymmetry, from Mirror Reflections to Superstrings* (W H Freeman & Co., 1991).

17

Identity

So far we have said very little about quantum mechanics, the other twentieth-century revolution in physics. (String theory is a quantum theory, but we haven't really used many of its peculiarly quantum features.) The next two chapters do address an issue that arises because of quantum mechanics. There are many books dealing with the subject and its puzzles for a popular audience, so I have decided to focus on an issue that is rarely found elsewhere: 'identity'.

There's a rich philosophical literature on this topic quite independent of quantum mechanics: why is one thing the same thing at different times, and why is one thing distinct from another? Such questions of course can have practical implications; part of the idea of temporary insanity as a legal defense is that at the time of the crime the person was literally 'not himself'. Or again, property rights require that property be reidentifiable over time; and if a painting is repeatedly touched up, is there a point at which it is no longer authentic? Other questions just seem perplexing: we can identify a wave as it moves across the lake, even though waves simply involve the vertical motion of water molecules. But what about when a wave meets another head on? We had two waves approaching each other, and then two waves moving apart, but who is to say whether they passed through each other or rebounded? (See figure 17.1.)

Or consider you and me; if we swapped what we are wearing, then clearly it would be you wearing my clothes and me wearing yours. But what if we swapped bodies, as happens in films sometimes? The movies (and philosophers) usually have it that it is you in my body and me in yours. What then if we swapped minds (and souls and every other property we may have)? Then I think our intuitions are torn: perhaps there is some 'me-ness' and 'you-ness' of us separate from all our properties, so that the swap results with me precisely like you before the swap, and you precisely like me before. (In the Middle Ages philosophers considered just such a feature, the 'this-ness', or *haecceity* in Latin, of things.) Or perhaps we have no 'identities' other than our properties, so the result of the 'total swap' is no change at all, just me and you exactly as before.

Philosophers of physics are often drawn to their subject because it allows precise, scientific(like) investigations of philosophical questions. So in this chapter and the next we will not look at anything as complicated

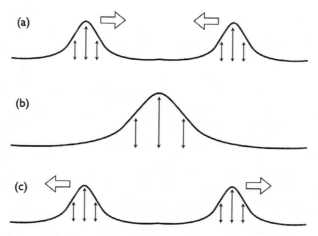

Figure 17.1 When a wave moves, water molecules move only up and down (like a wave on a rope). So when (a)–(c) two waves collide, do they rebound or pass through each other?

as the identity of persons; we will see that even for the very simplest kinds of things—elementary particles—these issues are rich and interesting. We won't even have space to look at particle identity over time; we will be absorbed by the question of whether and how things change when particles swap all their properties, and what this shows about their identities.

17.1 PARTICLE STATISTICS

One of the great triumphs of nineteenth-century physics—in the remarkable autumn of the era of 'classical' or 'Newtonian' mechanics—was to explain the gross behaviors of large collections of particles, especially how the temperatures and pressures and so on of gasses vary over time. The core idea, itself old, was that such behaviors can be understood in terms of the average or 'statistical' behaviors of their constituent particles. For instance, the temperature is a measure of the average kinetic energy (the energy due to motion). Other quantities are understood similarly, so the state of the whole—its collection of properties at a time—is derived from averages of the states of the parts—the properties of its constituent molecules.

Working out the statistical behavior requires knowledge of what the possible states of the collection of particles are. In particular, from the point of view of this discussion, the important question is: *given the number of particles and given the number of different states that each may possess individually, how many different states may the collection as a whole possess?* Our goal will be the philosopher's one of canvassing different possible answers and their implications for identity, not the physicist's one of calculating specific gross behaviors. Still, we will sketch some different behaviors that follow from different statistics, to emphasize that different state-counting rules make a difference to the observable behaviors of different kinds of particles.

First suppose (as we always will) that all the particles in a given collection are *identical*, in a precise sense used by physicists. This means two things: First, identical particles share the same constant properties by which we distinguish different kinds of particle; they all have the same masses, charges, and so on. Second, identical particles have the same possible individual, *single-particle* states. Especially, the location and motion of a particle typically is at least part of its state; then, for instance, the possible single-particle states of a particle confined in a box correspond to any position in the box and any motion whatever.

Suppose then that one has a collection of identical particles, say molecules of a gas in a box. Classical statistics are based on the assumption that a unique collective state is specified by specifying for each of the particles what its single-particle state is. It follows—and this is the crucial feature of classical particles in this discussion—that if two particles with distinct single-particle states were to swap their states, then a new, distinct collective state would obtain.

For instance, consider first a collective state in which the 3rd particle is stationary in the left side of the box and the 82nd is in the center moving vertically upward. Now consider the collective state in which all particles except the 3rd and 82nd are as before, and these have swapped states: now the 82nd particle is at rest to the left and the 3rd is in the center moving up. (See figure 17.2.) These two states are very similar—they agree on what points are occupied by particles and on the motion of the particle at any point—but according to classical state counting, they count as two distinct possibilities. This counting rule and the statistics

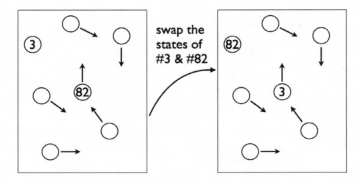

Figure 17.2 In Maxwell-Boltzmann (MB) counting, these two states are different, though their only difference is that molecules 3 and 82 have swapped states

that derive from it are named for two of the founders of classical statistical mechanics: 'Maxwell-Boltzmann' (MB) counting.

In the twentieth century it was discovered that other kinds of statistics—different state-counting rules—were needed to describe the gross behaviors of collections of *quantum* particles. 'Bose-Einstein' (BE) statistics describe photons and many atoms, and 'Fermi-Dirac' (FD) statistics describe electrons and other elementary atomic particles. For instance, at very low temperatures (a few billionths of a degree above absolute zero) BE particles form a 'Bose-Einstein condensate', in which they are all exactly alike in every way. And the electrons in atoms have the structure that they do because the 'Fermi Exclusion Principle' forbids any two fermions from being exactly the same; hence they cannot all orbit a nucleus in the same way, a fact that is crucial for understanding why different atoms have the different properties that they do.

The important point to make is that not only are different kinds of particles differentiated by their constant properties—electrons and photons have different masses, for example—but they may also be differentiated by their statistical properties: a given number of classical particles or electrons or photons with the same number of possible single-particle states will have different numbers of possible collective states. Thus we can classify types of particles according to their statistics, by their state-counting rules. So far we have seen MB or 'classical' particles, FD particles or 'fermions' and BE particles or 'bosons'. In a (partial) taxonomy its type of statistics defines a particle's genus, and its constant properties its species: for instance, while electron and proton are distinct species because they differ in mass and charge, they both fall under the genus of fermion.

17.2 SCHRÖDINGER'S COUNTING GAMES

Our first task is to investigate further these three kinds of counting and the possible philosophical differences between classical, MB, counting and quantum, BE and FD, counting; since it is experiment that determines counting rules, here we have a clear case of the direct relevance of experiment for philosophy. To help us we will draw on some beautiful analogies to simple counting games given by the physicist Erwin Schrödinger in a popular lecture in 1954.

In each game we are to imagine a teacher handing out awards to students, and consider how the nature of the awards affects the number of ways that he can do so. Each game corresponds to a type of particle with its characteristic statistics. More specifically, awards represent particles, and students represent possible single-particle states, so when a student is given some award, this represents the assignment of a particle to a state. (One way to get confused in this discussion is to slip into thinking incorrectly that students represent particles, so be careful.) Then the

Table 17.1 Maxwell-Boltzmann counting.

Curie	Noether	Poincaré	Weyl	Bill	Jill
Bill	Bill	Bill	Bill	4	0
Bill	Bill	Bill	Jill	3	1
Bill	Bill	Jill	Bill	3	1
Bill	Bill	Jill	Jill	2	2
Bill	Jill	Bill	Bill	3	1
Bill	Jill	Bill	Jill	2	2
Bill	Jill	Jill	Bill	2	2
Bill	Jill	Jill	Jill	1	3
Jill	Bill	Bill	Bill	3	1
Jill	Bill	Bill	Jill	2	2
Jill	Bill	Jill	Bill	2	2
Jill	Bill	Jill	Jill	1	3
Jill	Jill	Bill	Bill	2	2
Jill	Jill	Bill	Jill	1	3
Jill	Jill	Jill	Bill	1	3
Jill	Jill	Jill	Jill	0	4

games are constructed so that the number of ways of giving the awards to the students is equal to the number of distinct collective states for a certain statistical genus of particle. In Schrödinger's games each way of handing out the awards corresponds naturally to a distinct collective state of the corresponding particles.

- Consider first his game for classical, MB particles. Suppose that the awards, representing particles, are medals, each bearing the portrait of a different great scientist. The point about the medals is that they are distinguished by the portraits, so that it makes a difference which student gets which one: for instance, we have a different state of affairs if 'Mme Curie' is given to Bill and 'Poincaré' to Jill than if 'Poincaré' goes to Bill and 'Curie' to Jill (this situation is just like the case above, where we swapped the states of the 3rd and 82nd particles in a gas).

Suppose, for instance, that there are two students (in analogy to two possible states of each particle), Bill and Jill, and four medals (four particles), each bearing one of the busts of Curie, Poincaré, Noether, or Weyl. Each medal can go to either student, so for each medal there are two possibilities. And since there are four medals, as table 17.1 shows, there are a total of $2 \times 2 \times 2 \times 2 = 16$ possible ways of giving out the awards. For instance, Bill could receive all the medals, or Jill could, or Jill could receive all but Poincaré or all but Curie, and so on. The reader can easily check that there are six ways for Jill to receive two medals.

Now, it is vital to appreciate that in the analogy the different busts on the medals do not represent different *measurable* properties of the particles. Busts are fixed properties of the medals that one could in principle look at to tell the medals apart. However, there are no corresponding properties of the particles, since they are identical. As has been said, any measurement of a constant property of any of the particles will yield exactly the same result: the same mass, charge, 'spin', 'color' (to mention a couple of more exotic properties), or whatever. We don't count rearrangements of the single-particle states of classical particles as being distinct because they involve a measurably distinct state; since the particles are identical, no measurement can tell one state from another that differs only in that the 3rd and 82nd particles have swapped states. We count them as distinct because then we get the correct gross physical properties for collections of classical particles.

The busts must be similarly understood in the analogy: not as representing some observable property that could be used to identify particles empirically, but as 'labels' that function only to distinguish rearrangements. Thus we have a somewhat unfortunate, but well-entrenched terminology according to which MB particles, and analogously the medals that represent them, are simultaneously identical and 'distinguishable'. While they are identical in the physicist's sense, their states are counted as if they possessed labels which, like the busts, mean that it can make a difference which particle is in which single-particle state.

- To illustrate BE statistics, Schrödinger offers the following game (with a few liberties taken). Let each award, representing a boson, now be $10, deposited in the student's savings account. Since the net result of 'swapping' $10 from one student's account to that of another is no change at all, financial awards (and BE particles) are just not distinguishable in the way that medals (and MB particles) are, but are 'indistinguishable'. Since exchanges now make no difference, there are drastically fewer ways of handing out awards.

For instance, table 17.2(a) shows that there are only five ways to distribute $40 to two students: Jill gets either 0, 10, 20, 30, or 40 dollars in her account, and Bill gets the remainder. We had the same possibilities for medals, but because medals are distinguishable, it made a difference, say, which two medals Jill got. Since $10 in the bank is indistinguishable from any other $10 in the bank, there is now, for instance, only one way rather than six for Jill to receive two awards, only one way to receive $20. Analogously, the states of a collection of bosons differ only in how many bosons are in each single-particle state.

- For *fermions* (FD particles), let the awards be memberships in the Britney Spears fan club costing $10. Once again, students cannot swap their awards meaningfully: if Bill and Jill tried to swap their memberships, nothing would change, since they would both remain members. The point is that membership is a yes–no thing: there are

Table 17.2(a) Five ways of awarding $4 \times \$10$ bank deposits to two students, according to the rules of BE counting

Bill	Jill
$40	$0
$30	$10
$20	$20
$10	$30
$0	$40

Table 17.2(b) Five ways of awarding $4 \times \$10$ memberships to five students, according to the rules of FD counting

Al	Bill	Carl	Dee	Elle
✓	✓	✓	✓	-
✓	✓	✓	-	✓
✓	✓	-	✓	✓
✓	-	✓	✓	✓
-	✓	✓	✓	✓

not (in this club anyway) different kinds of membership. Thus memberships and fermions, like bank deposits and bosons, are indistinguishable: it makes a difference only how many memberships each student receives, or how many fermions are in each single-particle state (so memberships in this club aren't distinguished by different numbers or distinct identification cards or anything like that). But a new fact about memberships is that no one can have more than one: one cannot be a member twice over, since once you are in, you are in.

So if we have only two students but four memberships, there is no way to distribute the awards at all: there are zero possibilities. If there are exactly four students, then there will be just one way to award the memberships: one each. If there are five students, there will be five award possibilities, depending on which of the five students receives nothing (see table 17.2(b)). And so on.

This new fact about memberships represents the 'Fermi exclusion principle', mentioned earlier. It is important to understand that the principle does not arise from some new force that affects fermions and might affect other kinds of particle too. Instead it is an immediate and unavoidable consequence of what fermions are that the principle should hold; the principle holds necessarily in virtue of their being fermions, not in virtue of some external influence that might be turned off even as a conceptual possibility. The next chapter will have more to say about the principle. Schrödinger sums up his models thus:

> Notice that the counting is natural, logical, and indisputable in every case. It is uniquely determined by the nature of the objects.... Memorial coins are individuals distinguished from one another. Dollars, for all intents and purposes, are not, but they are still capable of being owned in the plural.... There is no point in two boys exchanging their dollars. It does change the situation, however, if one boy gives up his dollars to another. With memberships, neither has a meaning. You can either belong...or not. You cannot belong...twice over. (Actually, he used 'shillings' not 'dollars'.)

He is careful to emphasize that the games primarily illustrate the different systems of counting, and only secondarily the metaphysics of particles: one has to be careful about how literally Schrödinger thinks these models represent the reality of particle individuality. However, he is clear on three points. First, it is the 'natures' of the awards that determines how they can be handed out. In analogy, it's not just the exclusion principle that is essential to what it is to be a fermion; nothing about the counting rules could be switched off even in principle. This point is very nicely illustrated in these games; the analogy will be stretched a little in the new games we shall introduce to explain new kinds of quantum particles.

Second, Schrödinger emphasizes that the difference between these classical and quantum particles/awards lies primarily in the distinguishability of the former and the indistinguishability of the latter, in whether swaps make a difference or leave everything the same. Since the contrast is between one kind of classical particle and two kinds of quantum particle, we'll use 'quanta' to refer to bosons and fermions (and only these kinds of quantum particles, not any of the others introduced later). Then the point is nicely summed up by our saying that quanta, unlike classical particles, can only be aggregated: in the games for quanta, exchanges of awards leave things unchanged, so all one can do is count up how many dollars or memberships go to each student (correspondingly, how many particles are in each state—no limit for bosons, but at most one for fermions).

Finally, Schrödinger claims that all games that reproduce the statistics of quanta are of this kind, and that it follows that they are devoid of 'identity'.

> And this is just the salient point: the actual statistical behavior of [quanta] cannot be illustrated by any simile that represents them by identifiable things. That is why it follows from their actual statistical behavior that they are not identifiable things.

For Schrödinger, being an 'identifiable thing' means having some property of 'sameness' or 'identity' over and above any physically ascertainable properties. And of course in the games, the different busts on the medals stand for just such a kind of thing: the medals represent identical particles, so the busts correspond not to measurable differences but rather to properties that serve to distinguish for counting purposes. On the other hand, bank deposits and memberships—and any other analog of quanta, according to Schrödinger—do not have such properties: they do not differ from one another in any way, directly measurable or not.

Then, inferring by means of the analogy from the games to the kinds of particles, the different busts on the medals correspond to the different identities of the classical particles they represent, and the lack of distinguishing properties for bank deposits and memberships corresponds to the lack of different identities for the quanta that they represent. Classical particles are and quanta are not, according to Schrödinger, 'identifiable things'.

However, what we will see in the next chapter is that by focusing on only a part of the quantum formalism for many particles (that for quanta, the only part developed at the time), Schrödinger's conclusions oversimplify the issues surrounding identity and suggest a mistaken view about quantum particles. In particular, it is just not the case that the only significant difference between classical and quantum counting lies in distinguishability.

Schrödinger does not make quite this claim, since he also invokes the exclusion principle as a difference between MB particles and fermions. However, if one considers only quanta, it is much easier to say something revealing about indistinguishability than about exclusion, so Schrödinger was naturally led to emphasize the former in his analysis. A better target for this complaint is Paul Teller, who argues for a conception of quanta simply as indistinguishable entities, and is thus faced with the following question: "Given the intuitive picture of quanta that I have suggested, one expects that quanta of one kind [i.e., in some given state] can be aggregated without limit. Why, for Fermions, is there a limit of one to a kind? I have no answer to suggest."

The problem is that if classical mechanics simply meant one thing for individuality and quantum mechanics another, then we intuitively expect there to be not two but only one kind of quantum statistics, corresponding to the one kind of classical statistics. Why should there be one kind of quantum counting in which as many particles as you like can be aggregated in the same state and another in which at most one particle can be in a given state? What does this restriction mean?

I plan to show how taking a broader view of what kind of particles quantum mechanics allows gives a better understanding of particle individuality, indistinguishability, and exclusion. Quantum mechanics has room for particles other than quanta, including consistent non-classical, quantum statistics for *distinguishable* particles. Admittedly, no known elementary particles obey such statistics; all are quanta. And admittedly, quanta and classical particles do differ in regard to distinguishability, so one might suggest that this point is of formal relevance only.

But not so. The difference between classical particles and these new kinds of quantum particle can be traced to another aspect of identity, 'essential independence', and its converse, 'essential interdependence'. Thus we have a picture in which one can move from classical to quantum particles by modifying the notion of identity in either (or both) of two ways. And, we'll see, quanta are very special in this scheme of things: for them alone, indistinguishability is a consequence of their essential interdependence. So, even in the case of quanta, essential interdependence is relevant to understanding the metaphysics of individuality. And while Schrödinger says nothing strictly incompatible with these points, his silence on them means he only gives a part of the story.

Further Readings

Schrödinger's exemplary essay "What Is an Elementary Particle?" first appeared in the journal *Endeavor* in 1950, but it is probably most accessible as a reprint in his book of essays, *Science, Theory, and Man* (Dover Publications, 1957). The passages quoted come from 214–216.

In Paul Teller's *An Interpretive Introduction to Quantum Field Theory* (Princeton University Press, 1995), the first chapter is most relevant to this discussion.

You can find out more about Bose-Einstein condensation (and many other things) explained in a very easy-going way, with interactive demonstrations, on the *Physics 2000* Web site of the University of Colorado Physics Department: http://www.colorado.edu/physics/2000/index.pl.

Finally, a nice book for engaging these issues further and starting to explore their connections to other philosophical work is Peter Pesic's *Seeing Double: Shared Identities in Physics, Philosophy, and Literature* (MIT Press, 2003).

18

Quarticles

Twenty years after Schrödinger's lecture, physicists explored further theoretical, mathematical possibilities for quantum particles than being quanta. In particular, in the 1960s the physicists James Hartle, Robert Stolt, and John Taylor (HST) discovered a way of classifying the possibilities; bosons and fermions are in fact only two possibilities out of infinitely many in that scheme.

Their work is technical, and certainly the details are not necessary for our discussion. However, the general idea revolves around a familiar idea, that of a 'group'. Here, however, instead of displacements that move things around space, we have 'permutations' that swap particles among their states. Since the result is to change one collective state into another, if we think of the collection of collective states as a space—a 'state space'—we can think of the permutations as moving us from 'place' to 'place'. Then basically the HST classification scheme tells us what different series of swaps have the same effect, much as in geometry.

However, the scheme is only half the story, because they assumed that quantum particles are indistinguishable, but very recent work shows that in fact every kind of particle in the scheme, except quanta, comes in indistinguishable and distinguishable varieties (and varieties that might be described as being 'somewhat distinguishable', which we'll ignore). That is, one should visualize particles (even classical particles) as located as points in a two-dimensional classification: a particle's position along the first axis specifies what kind of particle it is in the HST scheme, while its position along the second axis describes whether it is distinguishable or not. Each such possibility corresponds to a unique kind of statistics and set of counting rules, and hence genus of particle according to the taxonomy introduced here earlier. There's no standard term to describe all these kinds of particles, so I will use *quarticle*.

18.1 NEW COUNTING GAMES

We'll investigate some of these other possibilities and their philosophical implications in the way we have so far, through some (Schrödinger-inspired) counting games. It turns out that Schrödinger has covered the easy cases, and the games for other kinds of quarticles are rather more complex. The two that we'll consider are the next simplest, but they already take a little thought. They represent the statistics for two kinds of quarticles that occupy the same place in the HST classification—their HST designation is '(1,1)'—but that differ in the other dimension: one is the distinguishable kind and the other the indistinguishable kind of (1,1) quarticle.

There is a precise technical reason for the name (1,1), but it's not important for our purposes, so to avoid technical jargon, here we'll call such quarticles *hookons*. (Actually there's a technical reason for this name too, but it should be a less distracting choice). (Like 'quarticle', this term is not current among physicists.)

- So consider first the game of *distinguishable hookons*. Imagine a fixed number of medals, all with different portraits and representing, as before, distinguishable quarticles, being awarded to students, representing states, as group prizes in two contests. The rules, stipulated by the alumnus whose bequest founded the contest, are a little baroque, but precise:

  (*i*)   The first contest can be won by any group that is no bigger than the number of medals, and the prize is one medal for each student in the group (students in the first group can also be in the second). However, the medals are not awarded individually, but are held in common by the group: in fact you can imagine the medals being displayed in the school trophy cabinet next to a plaque listing the members of the winning group.

  (*ii*) The remaining medals are the prize for the second contest and can be won by any group that is no bigger than the number of remaining medals (and that contains at least one student). Again the medals are held in common by the winning group, but this time different amounts of credit can go to the different members: each group member is assigned a whole number of the medals, and these numbers are recorded with the names on a plaque.

  (*iii*) Finally, under the terms of the bequest, there is one medal that must always be awarded in the first contest; any other can be awarded in either. And further, no one younger than the youngest winner of the first contest can be a winner in the second; no groups with such a member are eligible.

To illustrate, consider again the case in which there are four medals to award to two students (the analog of there being four hookons with two possible states each). Let the Poincaré medal be the one designated for

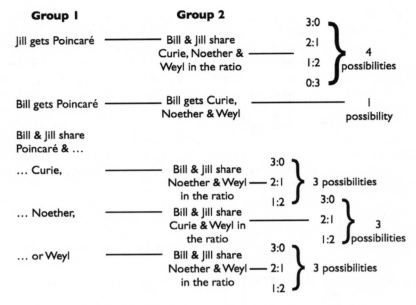

Figure 18.1 There are a total of $4 + 1 + 3 + 3 + 3 = 14$ ways of distributing three awards to two students according to the counting game for distinguishable hookons.

the first contest by rule (iii), and let Jill be the younger of the two. You can follow the state counting by referring to figure 18.1 as we go.

Since there are two students, the first group can contain either one or two students.

Suppose it contains only one, either Jill or Bill, who must receive the Poincaré medal. If it is Jill, then according to (ii) there are three medals to distribute to the second group, and the only choice is whether Jill receives three, two, one, or no shares in them (with Bill receiving the remainder), so there are four possibilities. But if Bill is the sole winner in the first group, then the age requirement of (iii) prohibits Jill from being a winner in the second group, so there is only one possibility: Bill is also the sole winner of the three remaining medals in the second group. Thus there are $4 + 1 = 5$ possibilities if there is only one winner in the first group.

What if there are two winners, Jill and Bill, in the first group? One of the medals they share is stipulated to be Poincaré, but then one of the three remaining medals must be selected for the second shared medal. And for each of these three possibilities, we have to decide how to distribute the last two medals to the second group; Jill can receive two, one, or no shares with Bill receiving the remainder (where receiving no awards means that the student is not in the group at all), for a further three possibilities. Thus there are $3 \times 3 = 9$ possibilities if there are two winners in the first group.

Hence there are $5 + 9 = 14$ possibilities for awarding four medals to two students in this game, a different number from the MB (16), BE (5), or FD (0) games, reflecting the different statistical properties of distinguishable hookons.

Let's talk about the meanings of the rules. Consider the first contest and rule (i): because the medals are awarded in common, there is no way two members of the winning group can swap their medals. While they are distinguishable—by the different busts on the medals—they behave as if they were indistinguishable within the winning group. Further, since there is one medal per group member, we have in effect the exclusion principle for the medals awarded to the group: there is no way for any student in the first group to get more than one share in the medals. That is, the medals given to the first group are rather like a collection of fermions.

What about the second group and rule (ii)? Clearly there is no exclusion principle, for it is explicitly envisioned that group members can be assigned more than one share in the medals. And of course how big each person's share is makes a difference when possible outcomes are counted. But once again in this group, since the medals are held in common, there is no way that two members can exchange their medals, even though they are distinguishable. Thus the awards given to the second group behave much like a collection of bosons.

So how does the distinguishability of the awards play a role? Because we do end up with a distinct arrangement of awards if one of the medals awarded to the first group is swapped with one of the medals awarded to the second group: different collections of medals are then held in common by the two groups. The awards behave like quanta only with respect to exchanges within one of the groups, not with exchanges between them.

Finally, let me draw attention to a new feature of this game. In the games that Schrödinger gave, remember, 'counting is natural, logical, and indisputable [and] uniquely determined by the nature of the objects' (1950, p. 214). The counting of distinguishable hookons is not similarly determined solely by specifying some classical analog—medals, bank deposits, or memberships—but also by specifying explicit rules concerning how they are to be distributed (in the earlier games the only rule was 'anything goes'). That is, the counting is not 'uniquely specified' by the nature of the awards, which are after all medals, like MB particles.

That this is so is unfortunate, because the quantum analogs of the rules are every bit as essential to distinguishable hookons as indistinguishability is to quanta (and exclusion to fermions), and the same is true of the rules that apply to any quarticles. Sadly, I don't think that there are any familiar entities whose natures would require them to be handed out in this way, and so the analogs are stuck with the rules. The reader will just have to imagine somehow that because of what they are, these medals could not be but awarded as they are; somehow the medals have the power to thwart other distributions (perhaps by irresistible mind control over the awards committee, or perhaps any other arrangement causes them to self-destruct explosively).

18.2 HOOKON IDENTITY

What lessons can be drawn from this game? First, one can have nonclassical, quantum statistics without indistinguishability. That is, the difference between classical particles and quantum quarticles, in general, is not that the former do and the latter do not possess identity in Schrödinger's sense. (Again, in all fairness, Schrödinger stops short of claiming that it is.) But if the difference between classical particles—MB quarticles—and distinguishable hookons does not lie in distinguishability, where does it lie, and how should we understand it?

In this game it is the stringent rules governing the awards that produces the nonquantum statistics, of course: for MB counting, 'anything goes' when medals are handed out to students, but in the last game there are all kinds of restrictions on who gets what. And in fact something analogous is true in the quantum systems these games represent. The rules of these games correspond to restrictions on the possible quantum mechanical states of the quarticles: which of all the mathematically possible states can quarticles of a given type actually possess?

The HST classification classifies quarticles according to the states they can possess: MB statistics arise (if the quarticles are distinguishable) if there are no restrictions at all, so all states are possible; while at the other end of the spectrum FD (and in a sense BE) statistics arise if the states are maximally restricted. Then other different kinds of quarticle statistics correspond to different state restrictions, and also to different rules in the Schrödinger-style games devised to represent them.

A technical discussion of the restrictions would be inappropriate for this book, but it is certainly possible to get an intuitive feel for them from the counting games, and in particular by considering one aspect of the hookon restrictions discussed earlier: the exclusion principle. Its effect is to prevent there being more than one medal (quarticle) per student (state) in the first group (the fermionic hookons). It makes the quarticles interdependent: whether an award can go to the first group depends on how many awards have already gone to it, and whether a quarticle can be in a given state depends on how many fermionic particles there are and what states have already been assigned to them.

Other rules impose the requirements that the students of the first and second groups be 'internally' indistinguishable. Corresponding state restrictions require that only states in which some quarticles are mutually fermionic or others mutually bosonic are allowed. But these kinds of restrictions also correspond to a quarticle essential interdependence when compared with MB particles. The rules again mean that whether a student can be in either group, and how many award shares she can obtain depend on what students have already been assigned prizes.

Thus the interdependence of quarticles constitutes a different way—from indistinguishability—in which the individuality of particles may be diminished by quantum mechanics, for their dependencies here are not dynamical: they do not concern how one quarticle will evolve given the states of the others, as do the dependencies of charged classical particles on one another. Instead they concern the states available to the quarticles, regardless of how they evolve (of course the dynamics cannot allow them to evolve out of the allowed states). That is, the restrictions concern the very ways of forming wholes that are possible.

Or in analogy again, the state restrictions are as much a part of the nature of the quarticles as the (in)distinguishability and exclusiveness are of the nature of medals, deposits, and memberships. Thus the way a quarticle can take its place in a collective state is dependent on what kind of quarticle it is and on what other quarticles are in the state: relative to classical particles, other kinds of quarticle fill states only collectively, not independently. And that is another sense in which they lack individuality.

Therefore, the messages of distinguishable hookons are, first, that quantum mechanics can mean either of two things for particle identity: one can move away from classical particles along a 'distinguishability' axis, or one can move along an 'interdependence' axis; and second, either way, one finds consequences for the individuality of the quarticles.

Now, two questions immediately present themselves. First, Schrödinger analyzed bosons and fermions in terms of indistinguishability (and exclusion only secondarily), but how does interdependence also play a role? Second, are these two axes independent? Specifically, is there in general, for any kind of restriction, both a distinguishable and an indistinguishable case? Let's take the questions in reverse order, and answer first for all quarticles but quanta.

18.3 INDISTINGUISHABLE QUARTICLES?

The answer in all cases, except quanta, is 'yes, there are distinguishable and indistinguishable kinds', a fact that can be illustrated for *'indistinguishable' hookons* (technically speaking, indistinguishable [1,1] quarticles) with another new counting game.

- This game is much the same as the last, except that it has as prizes both $10 club memberships and awards of $10 to be deposited into a student's savings account—both indistinguishable, of course—with a given total financial value of $10 \times n$. Then:

(*i*)   To the group that wins the first contest, which has at least one member (but no more than *n* members), we give each member a Britney Spears fan club membership worth $10.

(*ii*) The rest of money goes into the accounts of the students of the second group, with each member receiving a whole number (greater than zero) multiple of $10 in proportion to her contribution to the project.

(*iii*) As before, no one younger than the youngest winner of the first contest can be a winner of the second.

Thus every winning student will receive a prize of a certain dollar value, which may or may not include a membership. Because of what memberships are, it makes no difference if the members of the first group 'swap' memberships, and no more than one membership can be held. And because of the nature of money in the bank, it makes no difference if the members of the second group swap deposited dollars, as long as they keep the same amount of money each.

However, suppose that a winner in the first group, Anne, swapped her membership with the dollars of one of the winners in the second group, Zane; suppose, that is, that the result is also a possible way of distributing awards under the rules. Clearly the result (if possible) is distinct; in particular since Anne now has dollars and Zane now has a membership, it must be that they have 'swapped' groups to make an arrangement in which the groups are differently composed.

The reader can check that in our standard example in which the awards with a total value of $40 are awarded to two students there are eight distinct outcomes: more than for quanta but fewer than for distinguishable (1,1) and classical quarticles.

As the game suggests, indistinguishable hookons lack individuality in two ways. First, they are indistinguishable since they are represented by notes and memberships, whose natures mean that exchanges make no difference unless you exchange a note for a membership. Second, we have the hookon rules in place again (except that the indistinguishability of awards means we do not need to insist they are held in common), representing restrictions on their states. That is, relative to classical particles these quarticles have lost individuality both through indistinguishability and through interdependence. The same holds for all quarticles (including MB) except quanta, so distinguishability and independence do constitute independent dimensions of individuality, both of which should be taken into account.

18.4 QUANTA AS QUARTICLES

Quanta, however, are rather special, as bosons and fermions come only in an indistinguishable kind. Looked at from the point of view of our discussion, the essential interdependence of quanta forces them to be indistinguishable. Formally, the BE and FD state restrictions entail indistinguishability. This brings us back to the question about understanding quanta: what is missed if one analyzes their difference from classical particles in terms of distinguishability alone? Does the notion of interdependence shed further light on them?

There are two main points to address in answer to these questions. First there is Schrödinger's claim that 'electrons cannot be illustrated by any simile that represents them by identifiable things. That is why . . . they are not identifiable things.' I want to suggest a way in which Schrödinger is not entirely correct, which sheds light on how to think of quanta as quarticles. Second there is Teller's question of why there are two kinds of quanta. If the difference between classical and quantum is taken to be just the difference between distinguishability and indistinguishability, this state of affairs seems strange: now that we have added the dimension of interdependence to the discussion, can we give an answer?

Consider a final game which reproduces the counting of BE statistics but shows how interdependence forces indistinguishability (there's a similar game for FD statistics too, so my points hold for fermions as well). In this game the awards are—contrary to Schrödinger's claim— identifiable things, but their identities play no meaningful role, so they are indistinguishable. That is, it makes no sense for students to swap their awards.

The game involves medals, representing 'distinguishable' bosons, awarded in common to the best group, with at least one student and no more students than there are medals. Further, the students in the winning group may receive different amounts of credit for their work: a whole number proportion of the medals.

This game is familiar, of course, since it is the second half of the game played by distinguishable hookons. And similar results apply. It makes a difference which students are in the winning group, and it makes a difference how much credit each student gets. But since the medals are held in common, there is no sense in which the medals can be swapped among the winning students. So, for instance, with four medals and two students, there are five possible outcomes. (The game for fermions is the first of half of the game for distinguishable hookons: a certain number of medals to be held in common by the members of a group of the same number.)

As before, these rules force indistinguishability on erstwhile distinguishable awards: since they are held in common, rearranging is impossible. And something very analogous happens in the quantum formalism for bosons. Once we adopt the appropriate state restrictions, they have no choice but to be indistinguishable as far as any physical processes are concerned. Thus our considerations suggest a different way to think about bosons from that offered by Schrödinger. Now looking at them as just another kind of quarticle one sees them as differing from classical particles by their interdependence, though this happens to entail their indistinguishability. That is, we emphasize their difference from MB quarticles as a matter of the rules in the first place, and indistinguishability in the second place.

We can also use this game to clarify Schrödinger's claim that bosons are not identifiable things. Certainly this is the case in physical terms: we have just seen that the rules prevent the medals' identifying properties—the busts—having any consequences for counting. But perhaps, there are 'metaphysical' properties that outrun physics. Even classical particles are 'identifiable' things despite having having no measurable differences in their constant properties. One could say that bosons are individuals like classical particles or medals, that they have all the attributes that in some sense make them distinguishable, so that it is possible in some sense that they are in collective states in which swapping single-particle states makes a difference. However, the restrictions on which states are allowed—analogously the rules of the game for bosonic medals—prevent such states ever occurring.

Finally we can return to Teller's worry about the exclusion principle and the reason for two kinds of quanta, one arbitrarily aggregable and one restricted by the exclusion principle. If one thinks that the primary difference between classical particles and quanta lies in (in)distinguishability, then the existence of bosons *and* fermions is perplexing. But in this new picture we realize that there are not just two quantum statistics, but an infinity, corresponding to different points along two axes. In this picture bosons and fermions are the most strongly restricted—most interdependent—particles, and for the latter the state restriction implies the further restriction of exclusion, end of story. So, looked at as quanta—quarticles for which interdependence requires indistinguishability—fermions are more restricted than bosons, but if they are looked at as quarticles, both are highly restricted.

Further Readings

The ideas in this chapter come out of discussions with Tom Imbo about the technical work he has done with his graduate students on quarticles; credit for any original ideas here (and much of the mode of presentation) go to them. Here I have been able to report only some of the simpler results that they have found; they have also solved long-standing problems in calculating the physical consequences of different counting schemes.

19

Where Next?

So we've reached the end of our adventure. We've discussed a lot of the most important formal ideas in the philosopher of physics's tool belt: calculus, topology, geometry, mechanics, relativity, and particle statistics. Of course we have only skimmed the surface of these topics, but we have learned some important principles as well as seen what they are about. I've tried to give useful concepts and methods wherever possible rather than just pulling results out of hats.

We've also looked at a variety of philosophical approaches and techniques: the analysis of scientific concepts, the philosophy of language and metaphysics, logical construction, the exploration of the physical constraints on experience, and so on. And of course we've seen at length how these techniques allow the vital dialogue between physics and philosophy.

In this short conclusion I want to suggest where this material might point, both for the popular reader and for philosophy of physics.

The ideas and techniques discussed here should equip the reader to think intelligently about the philosophical implications of branches of physics we have not considered. For instance, I've recommended Brian Greene's *The Elegant Universe* several times as a source for understanding something of string theory. The reader of this book could read Greene's with a much improved background in the conceptual issues at stake and the philosophical implications of the theory. (For instance, in chapter 8 I mentioned the resurrection of conventionalism).

Moreover, we have seen how deep issues arise very quickly when one starts to think about some quite obvious facts about the physical universe: How is motion possible if every distance is infinitely divisible? Is space a something? Does it end? Is the present special? And so on. We've discussed quite a bit of the physics and philosophy needed to investigate these issues. But there are many more issues to think about: Could space be discrete, with a smallest distance between any two points? Why do we know more about the past than the future? Could a machine like a brain have free will? Could I be my own father if I had a time machine? Is there nothing to the properties of large bodies but the properties of their parts? The examples and ideas here should give you a good start in thinking intelligently about such questions.

If you are interested in studying philosophy of physics further, then the books listed at the end of each chapter are a good place to start. They all have more comprehensive bibliographies than I have attempted (I wanted to point to just a few of the very best sources for the beginner).

Let me mention one more book that might be of interest, since it comprises articles by leading philosophers of physics and physicists with philosophical interests (and me). It is *Foundations of Physics and Philosophy*, edited by Juan Ferret and John Symons (Automatic Press, 2009). Each essay explains what draws the author to foundational questions, how they understand the relevance of physics to philosophy and vice versa, and what they take to be the most promising areas for future research.

On that last question let me briefly say what I think. First there are still many foundational puzzles about quantum mechanics that need to be addressed. For instance, quarticles fail to be individuals in a familiar sense, so is there some better way of thinking about parts and wholes when it comes to quantum mechanics? If so, we might see that there is a deeper theory underneath being squeezed into an artificial form.

The puzzles are greatest in theories that try to give a quantum mechanical treatment of spacetime. String theory is the most popular approach with physicists, but there are others. It is likely that the successful theory will change our conception of space and time at least as much as relativity, and in totally different ways. It's popular to suggest that space and time are just appearances arising from a deeper reality, quite contrary to the picture that has dominated since Zeno at least: that space and time are the fundamental thing, which other things inhabit.

I've selected these topics for mention because they seem to offer the most promise of real interaction between physics and philosophy of the kind we have seen throughout the book. On the one hand, quantum theories, particularly of spacetime, are likely to undermine many of our ideas of reality and hence provoke a philosophical revolution. On the other hand, these areas in physics are crying out for the kind of analysis of fundamental concepts—perhaps more fundamental than those of space and time—that helped previous revolutions. The topics are technical enough that philosophers alone are unlikely to supply the key, but philosophers working with physicists could, I believe, help show the way.

Maybe you will be the philosopher who does.

Index